Sebastian Lenz

Stress – structure correlations in grafted polymer films

Sebastian Lenz

Stress – structure correlations in grafted polymer films

Phase transitions in constrained polymer films and why they are different from "normal" films and bulk

Südwestdeutscher Verlag für Hochschulschriften

Impressum / Imprint
Bibliografische Information der Deutschen Nationalbibliothek: Die Deutsche Nationalbibliothek verzeichnet diese Publikation in der Deutschen Nationalbibliografie; detaillierte bibliografische Daten sind im Internet über http://dnb.d-nb.de abrufbar.
Alle in diesem Buch genannten Marken und Produktnamen unterliegen warenzeichen-, marken- oder patentrechtlichem Schutz bzw. sind Warenzeichen oder eingetragene Warenzeichen der jeweiligen Inhaber. Die Wiedergabe von Marken, Produktnamen, Gebrauchsnamen, Handelsnamen, Warenbezeichnungen u.s.w. in diesem Werk berechtigt auch ohne besondere Kennzeichnung nicht zu der Annahme, dass solche Namen im Sinne der Warenzeichen- und Markenschutzgesetzgebung als frei zu betrachten wären und daher von jedermann benutzt werden dürften.

Bibliographic information published by the Deutsche Nationalbibliothek: The Deutsche Nationalbibliothek lists this publication in the Deutsche Nationalbibliografie; detailed bibliographic data are available in the Internet at http://dnb.d-nb.de.
Any brand names and product names mentioned in this book are subject to trademark, brand or patent protection and are trademarks or registered trademarks of their respective holders. The use of brand names, product names, common names, trade names, product descriptions etc. even without a particular marking in this work is in no way to be construed to mean that such names may be regarded as unrestricted in respect of trademark and brand protection legislation and could thus be used by anyone.

Verlag / Publisher:
Südwestdeutscher Verlag für Hochschulschriften
ist ein Imprint der / is a trademark of
OmniScriptum GmbH & Co. KG
Heinrich-Böcking-Str. 6-8, 66121 Saarbrücken, Deutschland / Germany
Email: info@svh-verlag.de

Herstellung: siehe letzte Seite /
Printed at: see last page
ISBN: 978-3-8381-1589-4

Zugl. / Approved by: Mainz, Max-Planck-Institut für Polymerforschung und Johannes Gutenberg-Universität, Diss., 2009

Copyright © 2010 OmniScriptum GmbH & Co. KG
Alle Rechte vorbehalten. / All rights reserved. Saarbrücken 2010

Contents

I	INTRODUCTION	1

II	FUNDAMENTALS AND METHODS	4

II.1 Stresses — 4
   II.1.1 Definition — 4
   II.1.2 Stresses and thermodynamics — 5
   II.1.3 Micromechanical cantilever bending phenomena — 6
   II.1.4 Micromechanical cantilever coating procedures — 7
   II.1.5 Micromechanical cantilever sensor arrays — 7
   II.1.6 Detection methods — 8
   II.1.7 NIR imaging interferometry — 10

II.2 X-ray and neutron experiments — 13
   II.2.1 X-ray reflectivity — 13
   II.2.2 X-ray and neutron reflectivity on polymer films — 16
   II.2.3 Roughness profiles — 18
   II.2.4 Correlated interfaces — 21
   II.2.5 Introduction to grazing incidence small angle x-ray scattering (GISAXS) — 24
      II.2.5.1 GISAXS geometry — 26
      II.2.5.2 The Distorted Wave Born Approximation — 27
   II.2.6 µ-x-ray reflectivity and µ-GISAXS on MC arrays at BW4 — 29
   II.2.7 Neutron reflectivity at N-REX$^+$ (FRM II) — 33

II.3 Thermodynamics of mixing — 37
   II.3.1 Free energy of mixing — 37
   II.3.2 Collapsed/stretched polymer brushes — 39
      II.3.2.1 Polymer brush: Definition and models — 39
      II.3.2.2 Collapse-stretching of polymer brushes in mixed solvents — 42
   II.3.3 Phase separating polymer blends — 43

II.4 Miscellaneous experimental techniques — 49
   II.4.1 Contact angle experiments — 49
   II.4.2 Gel Permeation Chromatography — 49
   II.4.3 Differential Scanning Calorimetry — 49
   II.4.4 Environmental scanning probe microscopy (SPM) — 50
   II.4.5 White light confocal microscopy — 50
   II.4.6 X-ray reflectivity from lab x-ray sources — 50

II.5 Substrate preparation — 52
   II.5.1 Substrate cleaning — 52
   II.5.2 Preparation of passivating Au films — 52

II.6 Materials — 53

III	GLOBAL SCATTERING FUNCTIONS: A TOOL FOR GISAXS ANALYSIS	55

III.1 Introduction — 55

III.2	**Theory**	**57**
III.2.1	Approximation of diffuse scattering by BA and intrinsic limits	57
III.2.2	Unified Exponential/Power-Law Fit model	59
III.2.3	Fractal objects	61
III.2.4	Weakly correlated systems	63
III.3	**Comparison with Simulations**	**65**
III.4	**Experimental verification**	**68**
III.4.1	Unified analysis from model systems	68
III.4.2	Unified analysis from novel TiO_2/(PEO)MA–PDMS–MA(PEO) films	71
III.5	**Summary**	**73**

IV THERMAL RESPONSE OF SURFACE GRAFTED TWO-DIMENSIONAL PS/PVME BLEND FILMS   74

IV.1	**PS/PVME bulk properties**	**74**
IV.2	**Thin film phase separation**	**77**
IV.3	**Preparation of grafted to polymer films**	**79**
IV.3.1	Introduction to specific and unspecific grafting to routes	79
IV.3.2	Surface functionalization with UV-sensitive benzophenone linkers	80
IV.3.3	Functionalization with polymers	82
IV.4	**Effect of grafting point densities**	**84**
IV.4.1	Hypothesis	85
IV.4.2	SPM results	85
IV.4.3	µ-XR and µ-GISAXS results	87
IV.4.4	Surface stress results	92
IV.4.5	Summary	96
IV.4.6	Grafting densities in fully active BP films	97

V STRESS/STRUCTURE CORRELATION IN GRAFTED FROM PMMA BRUSHES   103

V.1	**Motivation**	**103**
V.2	**Grafting from prepared polymer brushes**	**105**
V.2.1	Introduction to grafting from with atomic transfer radical polymerization (ATRP)	105
V.2.2	PMMA brushes prepared with surface initiated ATRP	106
V.2.3	Simultaneous MC sensor array/wafer coating	108
V.3	**Neutron reflectivity results**	**112**
V.3.1	Experimental objective	112
V.3.2	Data treatment	112
V.3.3	Fast collapse/swelling process	114
V.3.4	Gradual collapse/swelling transition	117
V.3.5	Summary of neutron reflectivity results	123
V.4	**Surface stress experiments**	**125**
V.4.1	Experimental approach	126
V.4.2	Stress propagation for the swelling of the collapsed brush of dry/swollen origin	127
V.4.3	Summary of surface stress results	129

| VI | SUMMARY AND OUTLOOK | 130 |

| VII | APPENDIX | 132 |

VII.1	Optical constants	132
VII.2	Dimensional and mechanic properties of Si MC sensors	133
VII.3	Input-files used for IsGISAXS simulations	133
	VII.3.1 Simulation of GISAXS from Au film	133
	VII.3.2 Simulation of GISAXS from TiO2 particles buried in a PMMA film matrix	135
	VII.3.3 Automatic IGOR Pro script for MC bending data analysis	136

| REFERENCES | 140 |

| PUBLICATIONS | 153 |

Meiner liebsten Anna und Ihrer roten Nase

I Introduction

Coated materials play an important role for everyday applications. Individual coatings can tailor the material's surface properties while maintaining the material's bulk properties, such as hardness, stiffness, toughness and so forth. Coated materials can have for instance adhesive[1, 2], lubricant[3] or repellent properties[4] towards fluids or different solid materials. Such properties allowed applications in the field of self-cleaning surfaces[5], sticky surfaces[2], anti oxidative surfaces[6], catalysts[7], paints[8] and adsorbents[9].

On large scale, coatings from inorganic materials can be deposited from their melt and with the use of electrochemical deposition techniques. On smaller scale, coatings can be applied using chemical vapor deposition or sputtering techniques[10-12]. Compared to inorganic coatings, polymer coatings can be prepared from more cost effective solution casting, such as dip-, blade- or spin-coating[13, 14]. Using defined drying processes, the quality of the individual coating can be controlled and the evaporating solvent can be recycled.

A second advantage of polymer coatings is that they can be adapted for individual applications. Designing individual polymer coated surfaces, the chemical property of the employed polymer can be varied for each functional coating application. However, the application of very specialized polymers may be not economical in some cases. It is more appropriate to create coatings for different functional purposes from polymeric materials, which are routinely prepared in large scale batches. Using different building blocks physical interactions between the single components can be used to tune the functionality of the individual coating. In such a way different film morphologies can be obtained[15] using block-co-polymers of different chain length aspect ratios. Consequently block-co-polymers can find use in various templating or nanoreactor applications[16, 17].

However, one major drawback of such physisorbed solution processed coatings, compared to inorganic coatings, is their usual low stability towards environmental conditions, such as continuous temperature changes or solvent exposure. One big aim for polymeric coatings in technical and everyday life applications is to make them stable against environmental conditions. This aim can be achieved by chemical polymer chain immobilization (also called grafting) onto the supporting bulk material. In contrast to non grafted chains, grafting changes the chains free energy. Free energy changes are accompanied with modifications of the physical coating properties. Thus different

Introduction

grafting routes can be used in order to tune the chains free energies and to obtain coatings with different physical properties.
Within this thesis two types of grafted polymer films were studied.

First, polymer chains were randomly grafted at more than one possible chain segment onto the substrate. Such grafting was achieved with the use of unselective UV sensitive silanizated benzophenone (BP) linkers[18]. The polymer chains were chemically immobilized to the functionalized substrate under UV exposure. The BP/polymer grafting point density could be adjusted tuning the BP's reactivity and surface coverage. Such changes in grafting point densities altered the chains free energies with accompanied different physical properties. Grafted polymer systems were composed of the poly-styrene (PS)/poly-vinyl-methyl-ether (PVME) blend and the two homopolymers. It is well known that the PS/PVME blend is macroscopically miscible at room temperature and phase separates at elevated temperatures[19-21]. Temperature dependent investigations with surface probe microscopy (SPM), µ-beam sized x-ray reflectivity (µ-XR) and µ-beam sized grazing incidence x-ray scattering (µ-GISAXS) on individual MC sensors[22, 23] gave insight in the polymer blends ability to phase separate. Quantitative domain size estimations allowed conclusions on the effect of grafting point density to the polymer chains free energy. The combination of structural investigations with surface stress investigations allowed to obtained detailed insight into the grafted polymer chain mechanics, which can be used for future developments using similar grafting routes.

In order to be able to analyze and quantify recorded GISAXS data, an analysis routine known for transmission SAXS analysis[24, 25] was adapted for GISAXS. The applicability including error estimations was discussed by theoretical considerations, which were supported with results from simulations and experiments. The pictured approach allowed quantifying structural domain's radii of gyration, forms and domains centre to centre distances. Quantifications within the pictured approach were of high importance for the explanation of phase separation mechanisms of the grafted films. In addition and to this date the analysis of GISAXS with the model's formalism helped to understand the charge transport mechanism of percolating networks and the formation of cadmium sulfide quantum dots.

Introduction

Second endgrafted (end tethered) polymer brushes were studied. It is well known that physical properties of polymer brush films can be tuned according to their chain lengths (polymeric weight) and grafting densities[26, 27]. High grafting densities force the polymer chains to stretch away from the substrate in order to minimize their free energy. Accordingly polymer brushes of high molecular weight will lead to thicker films, compared to low molecular polymer chains. Exposure of the polymer brush to a non mixing liquid leads to a collapsed brush phase, due to minor incorporations into the brush phase. In contrast, exposure to good mixing liquids leads to a swollen brush phase with higher film thicknesses, due to high incorporations into the brush phase.

Within this thesis the collapse/swelling transition of dense grafted poly-methyl-meth-acrylate (PMMA) polymer brushes, prepared with surface initiated atomic transfer radical chain polymerization (ATRP)[28, 29], were studied using mixtures of good and bad solvents. These transitions where analyzed with neutron reflectivity (NR), in combination with surface stress investigations using micromechanical cantilever (MC) sensors. Experimental results were compared with existing theories[30]. Combined results gave direct relations between the volume fraction of incorporated solvent, polymer/polymer and polymer/solvent interaction energies, adsorption/desorption phenomena and stress related chain kinetics.

Performed surface stress experiments have to be performed in the presence of liquid environments. Conventional surface stress experiments using deflection read out techniques[31] are not suitable for this kind of applications, because only relative MC bending data can be obtained. Thus, the initial bending of the MC sensor cannot be quantified. However, absolute quantifications of surface stresses are of extreme importance in order to understand PMMA brush kinetics. In addition the exchange of solvent is accompanied with a change in the refractive index and misaligns the focused laser spot. Thus no surface stress comparisons between two environmental solvent conditions can be drawn. To overcome this drawback a Michelson type interferometric setup was developed on the basis of a prototype setup, which allows to measure absolute surface stresses in liquid environments by recording MC sensor topographies.

II Fundamentals and Methods

II.1 Stresses

II.1.1 Definition

The term "stress" is originally used in engineering sciences. Stress describes the average distribution of a force, which acts per unit area on a deformable body. Thus, stresses are measures, which describe a body's reaction towards external forces, and are quantified in the units [Pa]. For a simple bar, which is elongated or compressed with a force ΔF along its centroid axis for ΔA, the resulting change in stress can be defined as

$$\Delta\sigma = \frac{\Delta F}{\Delta A} \qquad (\text{II.1.1})$$

The concept of "stress" can be also used, when the elastic mismatch of bilayered plates with two different biaxial elastic moduli is described, according to

$$E' = E(1-\nu) \qquad (\text{II.1.2})$$

where E is the elastic modulus and ν is the plates Poisson ratio.
In thin film approximations, where one of the plates (coating) is regarded to be < 5% in thickness than the second plate (substrate), resulting stresses can be approximated with Stoney's formula[32, 33]. Now the elongation of the thin coating is regarded along a one dimensional path and the resulting surface stress change is quantified in the units [N/m] according to

$$\Delta\sigma = \frac{E t_{Substrate}^2}{6(1-\nu)} \Delta\kappa \qquad (\text{II.1.3})$$

Stresses - Definition

$t_{Substrate}$ is the thickness of the substrate and $\Delta \kappa = \frac{1}{R_2} - \frac{1}{R_1}$ the change in curvature of a uniformly bent substrate with radius change of $\Delta R = R_2 - R_1$. The substrate reacts therefore with a bending upwards or downwards to lateral elongations or compressions of the film.

There are also other approximations in literature, which describe surface stress changes for biaxial plates with different thickness ratios[34-36]. However, within this thesis the studied systems are well in the range of $t_{coating} < 0.05 \, t_{substrate}$. Unlike explicitly mentioned, surface stresses are always calculated according to Stoney's formula.

II.1.2 Stresses and thermodynamics

The macroscopic mechanical engineering principle of surface stresses can be used to obtain physical insights in the microscopic interactions of films in nanoscale thickness ranges. For such applications, substrates - to which the resulting surface stresses are transduced - have to be thin and flexible enough for resolving stresses down to 1 - 10 mN/m. The resolution of such small surface stresses became possible by the use of micro mechanical cantilever (MC) beams, such as used in scanning probe microscopy (SPM). In contrast to SPM applications MC beams are used as sensing substrates for surface stress investigations.

With the help of bimetallic coatings, early applications ranged from heat sensors, gas sensors, nanoscaled spectrometers and calorimeters[37-39] towards studies on thiol and silane monolayer adsorptions [40, 41]. In these publications a deflection (bending) of the coated MC was detected. Butt[40] explained stresses resulting from proton adsorptions to SiO_x surfaces with changes of free surface energies. The free surface energy was defined as a normalization of the Gibbs free energy, in respect to the available surface sites. Recently Bergese et al.[42] were able to quantify free reaction enthalpies from DNA hybridization induced surface stress measurements. It is shown in this thesis that such kind of estimations can be also drawn for more complex adsorption/desorption phenomena, such as solvent adsorption in swelling polymer brushes. However, conclusions on free energy dependent polymer chain mechanics can only be drawn, if comparable information is available. For the studied polymer brush system comparative information, such as brush thickness,

Stresses
-
Stresses and thermodynamics

volume fraction of adsorbed solvent and polymer solvent interaction parameters was obtained from neutron reflectivity measurements.

In the case of surface grafted PS/PVME blend and homopolymer systems, attractive and repulsive interactions, resulting from free energy changes within the film system could be detected with MC bending experiments. Also in this complex film system, understanding of phase transition mechanisms was only possible with results from comparative x-ray reflectivity and grazing incidence small angle scattering experiments.

II.1.3 Micromechanical cantilever bending phenomena

As discussed above, MC bending results from surface stress changes in the functional coatings. Lateral acting attractive forces in the coating layer lead to MC sensor bending towards the side of the coating with $\Delta\sigma \propto \Delta\kappa > 0$ (corresponding to a tensile surface stress). Lateral repulsive forces lead to a bending of the NCS away from the coating with $\Delta\sigma \propto \Delta\kappa < 0$, corresponding to a compressive stress (Figure II.1).

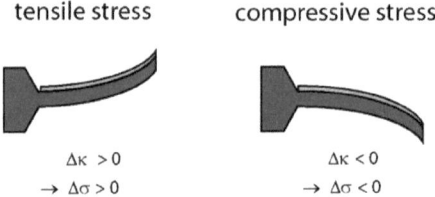

Figure II.1: Scheme denoting MC bending upon surface stress changes in the coating layer

However, forces within the coating layer can also act in the vertical direction in respect to the MC substrate. If resulting vertical stresses are transduced to the MC, the MC would bend into the opposite direction than for lateral acting stresses of the same magnitude. The effect on MC bending of lateral and vertical acting stresses can be estimated by the nature of the particular film system and from results from comparative experiments. Qualitative estimations on lateral or vertical surface stress changes could be drawn e.g. by measuring lateral domain size or film thickness changes.

II.1.4 Micromechanical cantilever coating procedures

As seen in Figure II.1 a selective coating of one MC side is of extreme importance in order to obtain reasonable results. Apart from some exceptions, the MC's topsides are coated with the functional layers. This is also done throughout this work, without exceptions. If both sides of the MC would be coated equally no bending signals would be obtained. There are several strategies how to obtain selective coatings of single MC sides[43, 44]. One is the use of a passivating layer on the opposite MC side, which can be removed after the coating process. Selective backside coating of the passivating layer can be achieved by single sided chemical vapor deposition or sputtering of suitable metal films. The functional layer can than be applied with e.g. dip coating or other techniques. However, there are technical drawbacks using such preparation routes. It has to be made sure that the passivating layer is removed quantitatively without altering the functional coating in the last step of sample preparation. Any residuals from the passivating layer on the MC backside or alterations in the functional layer can influence the bending signal. Especially, passivating residuals can influence temperature dependent measurements due to bimetallic effects. Alternative coating can be performed using solvent casting with the help of ink-jet techniques[45, 46]. Here no passivating backside coating of the MC is required. However, there are other drawbacks within this technique. One is a fast solvent evaporation in small droplet volumes, which limits its application e.g. in the case when polymer brushes are grown from the surface[47]. A second drawback can be patterning due to coffee stain effects occurring during droplet evaporation[48-50], which should be avoided for the formation of uniform films.

II.1.5 Micromechanical cantilever sensor arrays

For MC applications it turned out that single MCs exhibit baseline shifts by thermal drifts. Especially for sensing applications in liquid environments unspecific binding of analyte molecules not only to functional layers[51, 52], which carries the receptor molecules, but also on the backside of the MC was observed. To overcome these issues, arrays of MCs were designed[31]. The most commonly used MC array type consists of typically eight rectangular MC with lengths of 500 – 1000 μm, widths of 90 μm and thicknesses of 0.5 – 2 μm (Figure II.2). Using a combination of

Stresses - Detection methods

uncoated reference MCs and sensing MCs, which carry the functional coatings, it was possible to correct the obtained bending data for thermal drifts and unspecific adsorptions.

Figure II.2: Scanning electron micrograph of an MC array containing eight single MC sensors

Coating routines have to be adapted to the new sample geometry. Similar to the necessity of single sided MC coating it has to be made sure that only the desired sensing MCs are coated with the functional layer. In addition to plotting techniques[45, 46], MC arrays can be coated with the help of microfluidic networks[53] and microcapillaries[54-56]. It turned out that the plotting technique works reasonably well for the preparation of "grafted to" films, such as the demixing PS/PVME polymer blend system, studied within the framework of this thesis. However, for the preparation of "grafted from" polymer brushes such as the studied PMMA brush system, reasonable results are only obtained from reactor synthesis[57, 58]. Thus, for the preparation of "grafted from" films reference MCs have to be passivated with protecting metal films. The whole sample preparation route for the two systems is described in detail later in the thesis.

II.1.6 Detection methods

MC sensor arrays are mostly readout using beam deflection methods. Such methods are similar to SPM[59]. The laser is focused on the tip of the MC sensor and deflected to a detector. When the MC sensor bends the Laser spot on the detector changes its position, and the displacement of the MC can be calculated[40]. When the detecting instrument is built of eight equidistant arranged vertical-cavity surface-emitting lasers (VCSELs)) and Position Sensitive Detectors (PSDs), all MCs can be read out simultaneously[43].

Stresses - Detection methods

However using such deflection principles can cause several problems. First no information can be obtained on the original deflection of the MC before the experiment, because only changes in deflection can be measured. Second, changing refractive indices of the liquid environment, leads to shifts of the laser foci (Figure II.3), due to incident laser angles < 90°. Thus, the laser spot is most likely shifted to arbitrary positions and the device has to be recalibrated. Recalibration makes conclusions on surface stress changes upon refractive index changes, as by the change of a solvent, difficult. Especially, for solvent depending studies of polymer brushes no stress information can be obtained, when the refractive index of the solvent is changed upon solvent exchange. Prior brush swelling studies using deflection methods were therefore limited to solvent combinations of matching refractive indices[58].

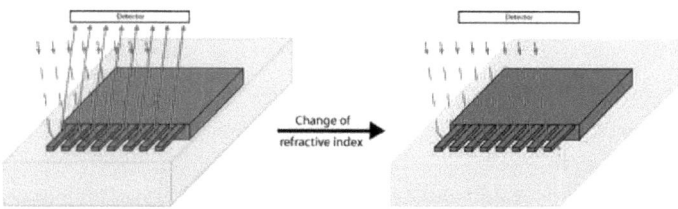

Figure II.3: Effect of change of environmental refractive indices on Laser foci.

To solve these issues one can use imaging interferometric techniques, which are capable of recording the whole MC arrays topography. Helm et al.[60] developed an interferometric white light phase shift technique, which is able to measure 3-D topographies of the MC sensors. The technique is based on a Michelson interferometric setup, which images the interference between light reflecting from the MC surface and reflections from a reference mirror. When the MC sensor is flat and not bent, the wave front gets reflected at equal z-position and the MC sensor image is superimposed by equidistant interference fringes. When the MC is bent the wave front gets reflected at different z-positions and the spacing between the interference fringes changes according to the curvature of the MC sensor. Phase shifts of interference signals lead to spatial shifts of interference fringe under conservation of spatial distances. From software analysis of multiple phase-shifted interference images accurate 3-D topographies are obtained.

Stresses
-
Detection methods

Compared to deflection methods, such absolute techniques have the advantages that they can determine the initial bending of MC sensors. Experimental interpretation is therefore not limited to relative MC bending data. However, changing environmental sample conditions from gaseous to liquid leads to the loss of interference fringes. The polychromatic light waves, which pass an optical medium, get dispersed with changing wavelengths and are therefore out of phase with waves reflected from the reference mirror. Thus, interferometric techniques using white light shift interferometry are limited to sensor application in gaseous environments.

II.1.7 NIR imaging interferometry

To overcome limitations discussed in the section above a near infrared (NIR) interferometric imaging technique was used. Here a coherent NIR laser of high wavelength stability is used. The wavelength is reduced passing an optical dense medium, such as glass or liquid and changes back, when leaving the medium. In such a way the reflected beam from the sample is still able to interfere with the reference beam.

An experimental interferometric setup, which was built during this work, is schematically drawn in Figure II.4. A laser diode with an operating power of 2.5 mW emits monochromatic laser light at λ = 785 nm. The divergent laser light is collimated with a collimator lens towards a biconvex lens (f = 40 mm), which focuses the laser beam to a pinhole (Φ = 5 µm). The filtered beam passes a lens (f = 50 mm) and is directed to a prismatic beam splitter, which transmits an intensity fraction to a piezo actuated mirror and reflects the remaining fraction to the sample cell. The MC array reflected laser light interferes with the laser light reflected from the reference mirror. Using a focal lens (f = 50 nm) the interference signal is displayed to a CCD camera and an interferometric image (Figure II.5) can be detected. Varying the position of the camera the amplification of the interferometric image can be adjusted. The commercial available OPTOCAT software (Breuckmann GmbH, Germany) was used to calculate 3-D topographies from multiple phase shifted interferometric images. Typically six images were recorded at different piezo voltages (10 – 29.5 V) in order to obtain 3-D topographies of high quality (Figure II.6). The pictured sample cell can be purged with gases and various liquids. A high temperature peltier element allows experimental temperatures from -10°C – 150°C.

Stresses
-
NIR imaging interferometry

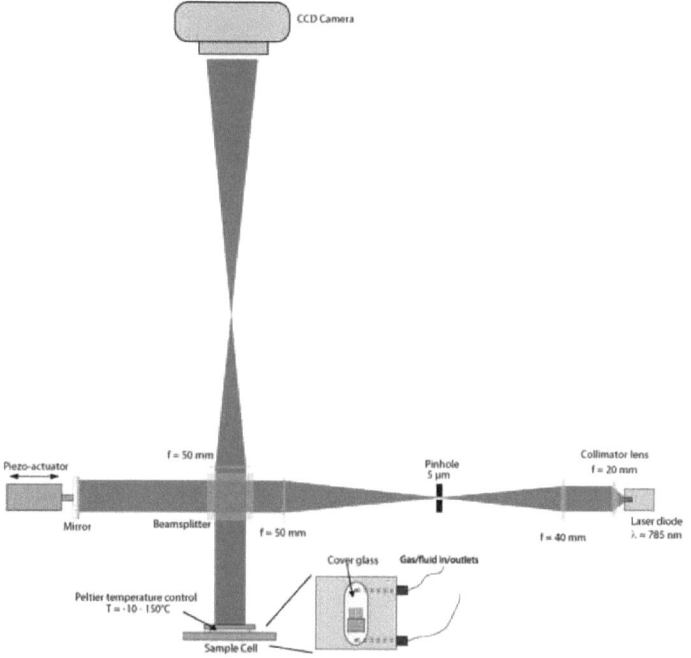

Figure II.4: Schematic representation of the built Michelson type interferometer device

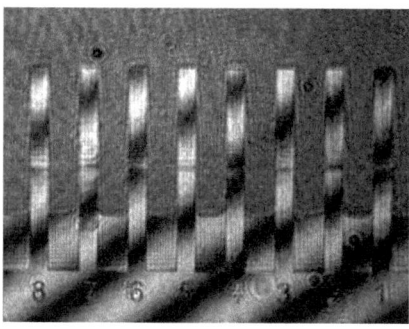

Figure II.5: Typical interferometric image of an MC sensor array containing eight MC sensors. In dark regions the laser beams interfere constructively. In bright regions the laser beams interfere destructively.

Stresses
-
NIR imaging interferometry

To obtain surface stress data, topography profiles along the MC sensors x-axes were performed. The obtained length/height data were modeled with a parabolic height profile[61] according to

$$z(x) = a_0 + a_1 x + \frac{a_2}{2} x^2 \qquad (\text{II}.1.4)$$

In Eq. (II.1.4) *z(x)* denotes the deflection of the MC sensor at each position x along the MC sensor. Under the assumption that the MC sensor is bent uniformly, the change of curvature $\Delta\kappa = \kappa_2 - \kappa_1$ can be calculated with

$$\kappa = \frac{d^2 z(x)}{dx^2} = a_2 \qquad (\text{II}.1.5)$$

For improved statistics typically six topography profiles per MC sensor were averaged. For data processing of typical > 5000 data files, which were extracted during one measurement run, a processing script was programmed for IGOR Pro 6.0 (Appendix VII.3.3). Thus, it was possible to obtain reliable e.g. surface stress vs. time graphs processed from ~ 7000 topography profiles extracted from ~150 3-D topographies.

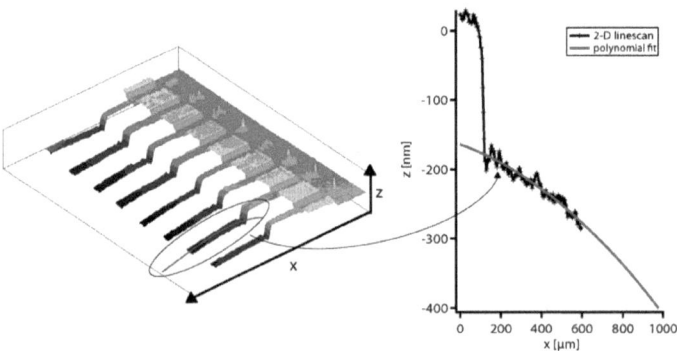

Figure II.6: MC sensor array's 3-D topography (left) with an extracted linescan (right). Curvatures were obtained from an approximation of the linescan based on 2nd order polynomials.

II.2 X-ray and neutron experiments

II.2.1 X-ray reflectivity

X-ray reflectivity is a tool to measure properties of thin films in a non destructive way. Using such techniques, film thicknesses, refractive indices and the film's roughness can be parameterized quantitatively[62, 63]. For less absorbing soft matter/polymer films accurate values can be obtained within thickness ranges of ~ 0.5 – 500 nm. An x-ray reflectivity experiment is conducted as following. A monochromatic x-ray beam is direct to the sample's surface with a certain incident angle α_i. The beam is transmitted and reflected at the interface and the reflected intensity is measured in respect to α_f. In conventional reflectivity or Bragg scans, α_i is steadily increased from 0.1 - 3° and the reflected intensity is measured at $\alpha_f = \alpha_i$. It has to be noted that experiments performed at higher α_i can give information on the films crystal lattice. However, such kinds of experiments were not performed for the analysis on amorphous polymer films. When a beam coming from an optical thinner medium is reflected at the interface to an optical thicker medium, under α_i smaller than the critical angle α_c, it will be totally external reflected. The obtained reflectivity is therefore equal to 1. At $\alpha_i > \alpha_c$ the reflectivity is decreasing with increasing α_i.
The beam's refraction can be expressed with Snell's law

$$n_0 \cos(\alpha_i) = n_1 \cos(\alpha_\tau) \qquad (\text{II.2.1})$$

For air, the refractive index n_0 is 1. Under total external reflection α_τ is 0 and α_c can be expressed and approximated for small angles with

$$\alpha_c = \arccos\left(n_1/n_0\right) \qquad (\text{II.2.2})$$

X-ray and neutron experiments
X-ray reflectivity

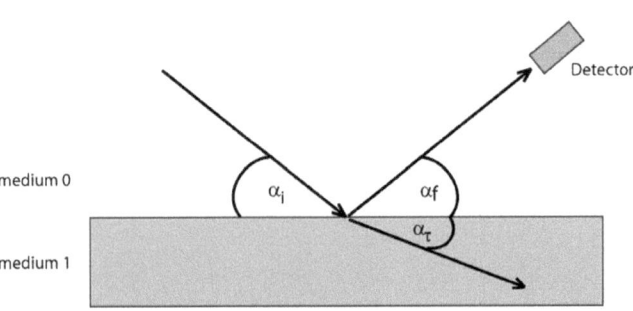

Figure II.7: Experimental reflectivity geometry; an incident beam is reflected at the sample towards the detector with ($\alpha_i = \alpha_f$).

For $\alpha_i > \alpha_c$ the reflected intensity $\overline{R}$ at an homogeneous interface is expressed by the product of the real and imaginary part of the reflected amplitude r

$$\overline{R} = r \cdot r^* = r^2 \qquad (\text{II.2.3})$$

With the equation for the reflected intensity expressed by the z component of the single wave vectors

$$r = \frac{k_{z0} - k_{z1}}{k_{z0} + k_{z1}} \qquad (\text{II.2.4})$$

one obtains *Fresnel's law of reflectivity*

$$\overline{R} = \left| \frac{k_{z0} - k_{z1}}{k_{z0} + k_{z1}} \right|^2 \qquad (\text{II.2.5})$$

A homogeneous flat one component film of a certain thickness t has usually one interface with its substrate and one interface with the surrounding environment (air/vacuum). An incoming beam is therefore reflected and transmitted at two interfaces.

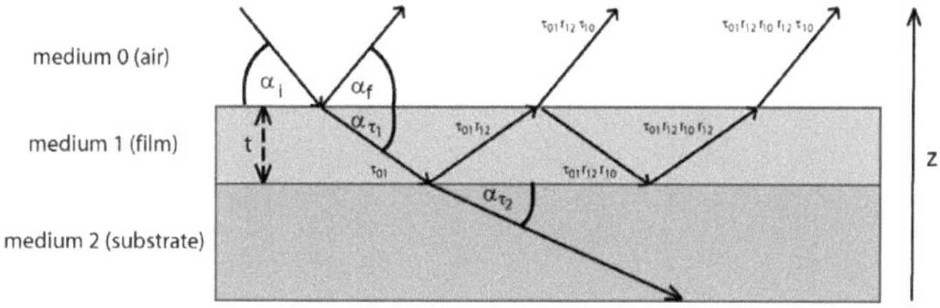

X-ray and neutron experiments
-
X-ray reflectivity

Figure II.8: Diagram of the beam path in a film sample, composed of a film medium with thickness t located on a substrate in surrounding air. The beam is reflected and transmitted at the air/film and film/substrate interfaces.

Multiple constructive and destructive interferences resulting from the transmitted and reflected beams as illustrated in Figure II.8 is derived as[62]

$$r = \frac{r_{0,1} + r_{1,2} \exp(2ik_{z,1}t)}{1 + r_{0,1}r_{1,2} \exp(2ik_{z,1}t)} \quad (\text{II}.2.6)$$

For the reflected intensity $\overline{R}$ can be written as

$$\overline{R} = \left| \frac{r_{0,1} + r_{1,2} \exp(2ik_{z,1}t)}{1 + r_{0,1}r_{1,2} \exp(2ik_{z,1}t)} \right|^2 \quad (\text{II}.2.7)$$

From Eq. (II.2.7) reflectivity profiles can be calculated and compared with experimental data. Figure III.3 shows a simulated reflectivity profile for a flat polystyrene (PS) film of $t = 20$ nm. From the width of the observed minima the film thickness t can be estimated by

$$t = \frac{2\pi}{\Delta q_z} \quad (\text{II}.2.8)$$

with $q_z = k_{z,0,f} - k_{z,0,i}$.

X-ray and neutron experiments
X-ray reflectivity

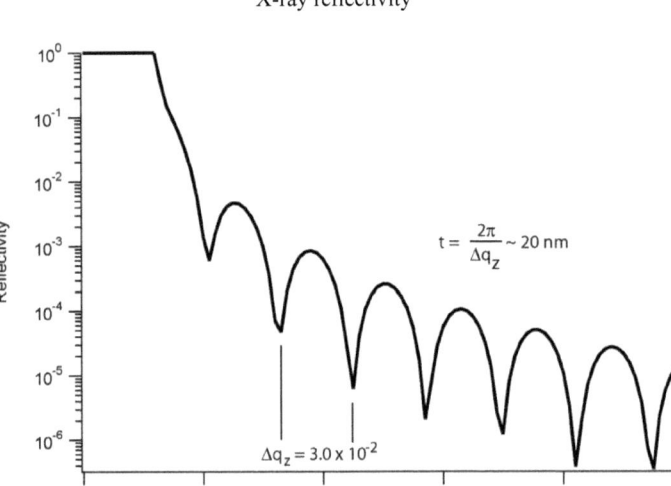

Figure III.3: Calculated reflectivity profile for a homgeneous flat polystyrene film with $t = 20\ nm$. The film thickness can be estimated by the position of the minima according to Eq. (II.2.8).

II.2.2 X-ray and neutron reflectivity on polymer films

X-ray and neutron reflectivity techniques are valuable tools to study polymer films[62, 64, 65]. The difference of these two techniques is the difference in the beam's character, leading to sensitivities towards different material properties. X-rays are of electromagnetic nature. Therefore they interact with the atoms electron shell. X-ray experiments are of high value, when materials with high contrasts in electron densities are studied. Such as in materials containing elements with high differences in atomic numbers Z.

The refraction of the incident beam can be explained using the refractive index

X-ray and neutron experiments

X-ray and neutron reflectivity on polymer films

$$n = 1 - \delta - i\beta \quad (\text{II.2.9})$$

The value δ accounts for the x-ray beam scattered at the atoms electron shell. It is described by the materials electron density ρ_{el}, the wavelength λ of the x-ray beam and the classical electron radius r_0 by

$$\delta_X = \frac{\lambda^2 \rho_{el} r_0}{2\pi} \quad (\text{II.2.10})$$

The complex term $i\beta$ describes the absorption of x-rays by matter, with a mass absorption coefficient μ by

$$\beta_X = \frac{\lambda \mu}{4\pi} \quad (\text{II.2.11})$$

Typical values for δ are in the order of 10^{-6}, while β is for polymer samples typically 2 to 3 orders of magnitude smaller. Thus the refractive index n_1 is near unity, and < 1.

Despite its high accuracy due to typical high beam intensity/background ratios in particular at synchrotron beamlines, the use of x-rays in reflectivity experiments is limited. There are two major drawbacks. First the strong interaction of x-ray waves with the electron shells, leads to low penetration depths. For water the penetration depth of x-rays with $\lambda = 1.54$ Å at $\alpha_i = 1°$ is less than 10 µm. The second drawback is that materials, such as polymer and solvents are of similar electron density. Thus, one obtains less refraction at such interfaces.

Neutrons are described as a particle wave. They interact with the atoms nuclei. Therefore high scattering contrasts are obtained within different isotopes of the same element. Especially, high contrasts are obtained by exchanging Hydrogen atoms with Deuterons. Thus isotopic Deuteron labeling of e.g. organic solvent molecules can lead to high refractive index contrasts. The second advantage is the high transparency of many elements towards neutrons. Since Neutrons interact with the atomic core, they can easily penetrate matter. The imaginary part can therefore except for some elements, e.g. cadmium, be neglected.

For non magnetic substances the refractive index for neutrons is given by Eq. (II.2.9) with the real part

$$\delta_N = \frac{\lambda^2}{2\pi} N_A \sum_i \frac{\rho_i}{m_i} b_i \qquad (\text{II.2.12})$$

where b_i is the neutron scattering length, ρ_i the density and m_i the molecular mass of component i. N_A is the Avogadro constant. In the case of polymer materials Eq. (II.2.12) can be simplified by

$$\delta_N = \frac{\lambda^2}{2\pi} N_A \rho \frac{b_{Mon}}{m_{Mon}} \qquad (\text{II.2.13})$$

where b_{Mon} is the scattering length of one repeating unit with mass m_{Mon} and density ρ.

One of the neutrons drawback compared to x-rays is their usual low flux, which leads to comparable poor intensity/background ratios. The second drawback is that large experimental facilities, such as nuclear reactors or spallation sources, are needed, which leads to high experimental costs.

II.2.3 Roughness profiles

The interface between two media is not ideal flat. Reflectivity can therefore usually only be explained in zero approximation by step like reflectivity models (Eq. (II.2.7)). Statistical height variations at the interface of two media have to be implemented in the reflectivity theory[62, 66]. Roughness can be distinguished in waviness and micro structured roughness. When the radius of a rough structure is higher than the coherence length of the probing beam, one speaks of waviness. In the other case, when the radius of the rough structure is smaller than the coherence length, the interface is rough.

On rough surfaces the film/air interface is out of phase with the substrate/film interface, which leads to decreases in constructive and destructive interferences of the reflected beam. As a result one will observe dampening of the observed oscillations in reflectivity profiles compared to perfectly flat interfaces.

X-ray and neutron experiments
-
Roughness profiles

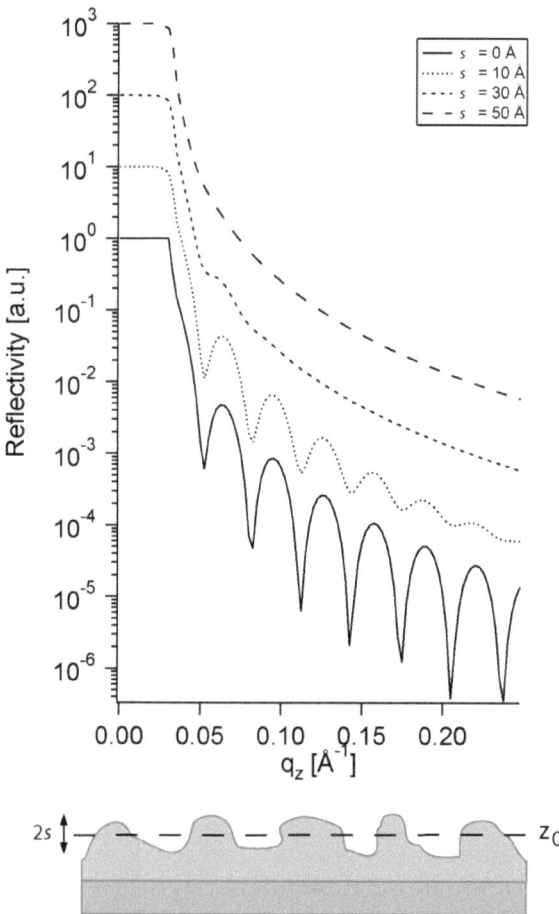

Figure II.9: Comparison of reflectivity profiles from perfectly flat and rough film surfaces with different σ and constant t = 20 nm. For a better illustration the reflectivity profiles are shifted by a factor of 10. The lower scheme illustrates a microstructured rough film placed on a flat substrate.

Such deviations $w(z)$ from an interface at position z_0 can be described under the assumption of statistical distributions with a Gaussian distribution with variance s^2

X-ray and neutron experiments
Roughness profiles

$$w(z) = \frac{1}{\sqrt{2\pi}s} \exp\left(-\frac{z-z_0}{2s^2}\right) \qquad (\text{II.2.14})$$

After integration one obtains the *error function*

$$n(z) = n_1 + \Delta n \frac{1}{\sqrt{2\pi}s} \int_{-\infty}^{z} \exp\left(-\frac{u^2}{2s^2}\right) du = n_1 + \Delta n \cdot erf(z) \qquad (\text{II.2.15})$$

The refractive index profile obtained from Eq. (II.2.15) can be simplified with the use of a *tanh –* like profile[67]

$$n(z) = n_1 + \frac{\Delta n}{2}\left(1 + \tanh\left(\frac{2z}{a_1}\right)\right) \qquad (\text{II.2.16})$$

When s and a_1 are scaled according to $\sqrt{2\pi}s = a_1$ such simplification is reasonable, due to the similar curve progression of an *error function* and the *tanh*.

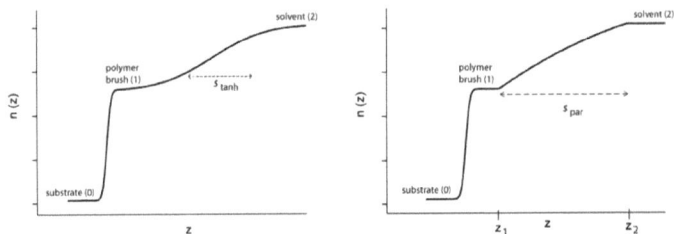

Figure II.10: Comparison of refractive index profiles of a substrate supported endgrafted polymer brush immersed in a good solvent. The curves were simulated according to refractive indices calculated for neutrons with λ = 4.26 Å. The right profile shows the obtained profile with a combination of *tanh* functions (II.2.16). The left profile was calculated for the parabolic formalism (II.2.17).

Apart from step function profiles with *err/tanh* roughness profiles, a number of other roughness models can be found in literature. One uses parabolic profiles for the description of density profiles of polymer brushes in good solvents (Eq. (II.2.17))[68, 69]. Details on the physical interpretation are discussed in chapter II.3.2.1.

X-ray and neutron experiments
-
Roughness profiles

$$n(z) = a + b \cdot z + c \cdot z^2 \qquad (\text{II}.2.17)$$

Parameters a, b and c have to be chosen for n being steadily at $z = 1$, $z = 2$ and dn/dz being steadily at $z = 0$. Figure II.10 shows a comparison of a simulated polymer brush refractive index profile simulated from a *tanh* type (II.2.16) and parabolic type (II.2.17) function. From the distance between the two turning points in the *tanh* profile the roughness s_{tanh} can be obtained. From the distance between z_2 and z_1 in the parabolic profile s_{par} can be obtained.
Comparison shows that absolute values are proportional by a factor of 2.

$$s_{par} \approx 2 s_{tanh} \qquad (\text{II}.2.18)$$

II.2.4 Correlated interfaces

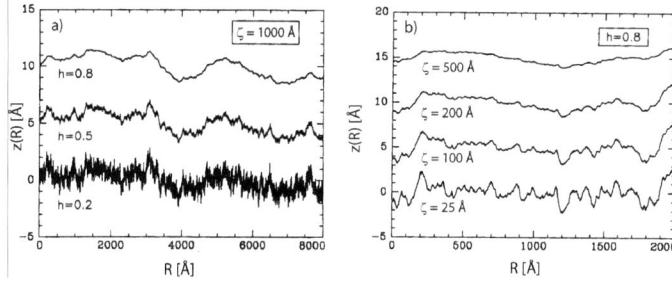

Figure II.11: Height profiles corresponding to the correlation function from Eq. (II.2.19); a) dependence of the height profile in respect to Hurst parameter h with constant s = 1 Å and ζ = 1000Å ; b) dependence of the height profiles in respect to correlation length ζ with constant s = 1 Å and h = 0.8;[i]

In addition to fractal self-affine structures found in nature[70], such as coastal lines, clouds and so on, many microstructured surfaces can be described with self-affine fractal models. Examples are capillary waves, thin films[71-73] and polymer brushes[74].
Sinha et al.[75] proposed a simple height/height correlation function to describe the self-affinity of

---
[i] adapted from Tolan, M., *X-ray scattering from soft-matter thin filmsmaterials science and basic research*. Springer: Berlin [u.a.], 1999.

X-ray and neutron experiments - Correlated interfaces

common isotropic surfaces:

$$C(R) = s^2 \exp\left\{-(R/\zeta)^{2h}\right\} \qquad (\text{II.2.19})$$

The correlation function describes the probability that two points on a surface separated by R have the same height position $z(R)$. Eq. (II.2.19) is a function of the Hurst parameter h and the lateral correlation length ζ. The Hurst parameter h, which is restricted to the range $0 \leq h \leq 1$, describes the smoothness of the film. Small values for h correspond to jagged surfaces, while values close to unity correspond to smooth surfaces (Figure II.11a). The correlation length ζ describes the lateral length scale, where the interface appears to be rough. For $R \ll \zeta$ the interface is self-affine rough, while for $R \gg \zeta$ the interface appears to be smooth (Figure II.11b).

However, the description of a rough interface by only one correlation function is not complete if the roughness from an interface such as a substrate is partly transferred to the following interface, such as a thin film or polymer brush/air interface (Figure II.12). According to Spiller et al.[76] the Fourier transform of the real space height profile of a lateral correlated $z_k(R)$ function can be expressed as a replicated form from interface j and an intrinsic part, which would be present without the other interface. Such a height profile function can be expressed in the reciprocal $q_\parallel$ space by

$$z_k(q_\parallel) = \chi_{jk}(q_\parallel) z_j(q_\parallel) + z_{k,\text{int }r.}(q_\parallel) \qquad (\text{II.2.20})$$

The reciprocal $q_\parallel$ plane is composed of the lateral q_x and q_y components of the scattering vector q by $\sqrt{q_x^2 + q_y^2}$. Thus scattered intensities in the $q_\parallel$ plane can be attributed to lateral correlations in real space. The full definition of $q_\parallel$ is given in chapter II.2.5.1. $\chi_{jk}(q_\parallel)$ is the replication factor, which describes how the Fourier of the height profile from interface j, $z_j(q_\parallel)$, is transferred to interface k. If $\chi_{jk}(q_\parallel)$ is close to unity the profile is perfectly replicated and only modified by the intrinsic part $z_{k,\text{int }r.}(q_\parallel)$. In the other case, when $\chi_{jk}(q_\parallel)$ is close to zero interface k is laterally uncorrelated to interface j, and only $z_{k,\text{int }r.}(q_\parallel)$ is left. The term $z_{k,\text{int }r.}(q_\parallel)$ can be derived for various film types[63]. However, analytical height profile descriptions where not performed on experimental data and are therefore not included in this chapter.

X-ray and neutron experiments
-
Correlated interfaces

In contrast to uncorrelated rough interfaces (Figure II.12b), height profiles of two partly correlated interfaces (Figure II.12a) are in phase. As a consequence enhancements of oscillations in reflectivity profiles obtained from partly correlated films compared to uncorrelated films with the same s value are observed. Fitting of experimental data from correlated rough films will therefore result in lowered s values, than actually present.

Plots of specular reflected intensity vs. q_z data disregard in plane scattering in the $q_{\|}$ direction, which accounts for lateral correlations. Two dimensional q_z vs. q_x intensity contour plots unravel off-specular scattering at $q_x \neq 0$[77-80]. The reciprocal q_x direction is related to experimental incident and exit angles according to:

$$q_x = \frac{2\pi}{\lambda}\left(\cos(a_f)\cos(2\theta) - \cos(\alpha_i)\right) \quad (\text{II}.2.21)$$

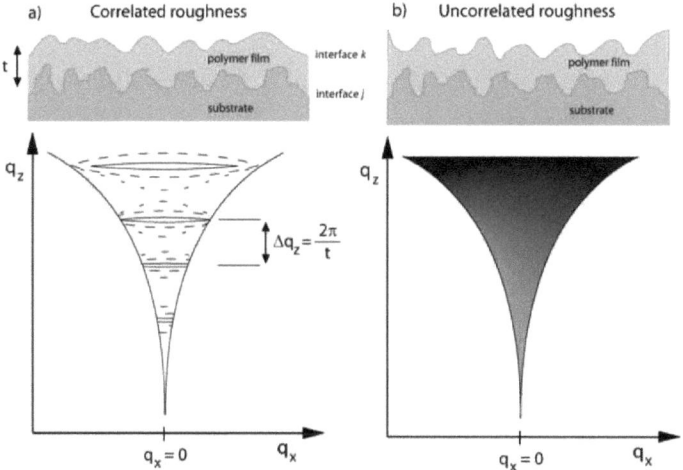

Figure II.12: a) Schematic presentation of a partly laterally correlated thin (polymer) film located on an intrinsic rough interface (substrate) with according schematic q_z vs q_x intensity contour plot; as illustrated the intensity drops with appearing oscillations in q_z; b) Schematic presentation of an uncorrelated rough film with equal s as in the correlated case; as illustrated below the intensity drops exponentially with q_z and no oscillations appear.

X-ray and neutron experiments
-
Correlated interfaces

For in plane off specular scattering the out of plane angle 2θ (Figure II.13) is set to zero. Contour plots are limited in q_x and q_z as indicated by the circle segments, due to an experimental limitation of $\alpha_i > 0$ and $\alpha_f > 0$. Such data can be obtained by either performing rocking scans with a 1-D or point detector or by analyzing the reflected beam profile with the use of 2-D detectors, according to α_i and α_f.

Contour plots for specular and off-specular reflected beam intensities from correlated rough interfaces show defined oscillating contours (Figure II.12a), which is a direct consequence for the resonating reflected beam. In contrast, one observes exponential intensity decays with q_z for uncorrelated rough interfaces. Within this work it was of interest to compare absolute s values from obtained neutron reflectivity experiments, irrespective of roughness correlations. Using contour plots it becomes possible to correct specular reflectivity data from correlated roughness and background contributions. In this purpose off specular profiles at constant $q_x \neq 0$ parallel to the specular reflection profile at $q_x = 0$ can be drawn and subtracted from the specular reflectivity profile. In such a way good corrections for correlated roughnesses and background noise are obtained.

II.2.5 Introduction to grazing incidence small angle x-ray scattering (GISAXS)

Grazing incidence small angle x-ray scattering (GISAXS) can probe structural film properties perpendicular and parallel to the sample plane. Information on film structures perpendicular to the sample plane is usually obtained by the analysis of the specular reflected beam in $q_\perp$ (q_x, q_z) direction as explained above. In addition lateral film structures can be studied by either analyzing scattering parallel to the sample's surface at the specular beam position or by analyzing additional off specular scattering. Using 2-D detectors and by integrating scattered intensities over time, typical GISAXS patterns can be obtained. Arbitrary intensity profiles in reciprocal $q_\perp$ and $q_\parallel$ spaces can be obtained after rebinning scattered intensities over a small number of detector pixels (typically < 10).

The aim in probing lateral length scales of colloidal and polymer composed films is to obtain structural information from a few nanometres up to hundreds of nanometres[81, 82]. Such structures

X-ray and neutron experiments

Introduction to grazing incidence small angle x-ray scattering (GISAXS)

can be located as islands at the film/air interface or buried within a thin film matrix. In the second case, penetration of the x-ray beam into the whole film is of high importance to obtain scattering data averaging structural attributes perpendicular to the sample plane. Therefore incident angles α_i, which are sufficiently higher than the materials critical angle α_c should be chosen. Using experimental setups with $\alpha_i > \alpha_c$, one can benefit from a splitting up of the reflected intensity into two peaks in $q_\perp$ (q_x, q_z) direction (Figure II.13). With the help of beryllium lenses[83, 84], equipped at high energy synchrotron source beam lines, incoming x-ray beams are focussed to small sample spots. As a consequence it became possible to obtain high scattering intensities in small scattering volumes at $\alpha_i > \alpha_c$.

One part of the splitted beam is the specular reflected beam at an angle of the exit beam α_f equal to α_i. The second is the material dependent Yoneda peak[85, 86] at $\alpha_f = \alpha_c$[87, 88]. Diffuse scattering at this peak position can be explained within the Distorted Wave Born Approximation (DWBA)[75]. For practical reasons a separation of the two peak positions is useful, because scattering at the specular beam position, $\alpha_f = \alpha_i$, can be separated from diffuse scattering occurring at the Yoneda peak, $\alpha_f = \alpha_c$[89]. Hence, deconvolution of the primary reflected beam profile, scattering at the specular peak position and diffuse scattering at the Yoneda peak can be avoided.

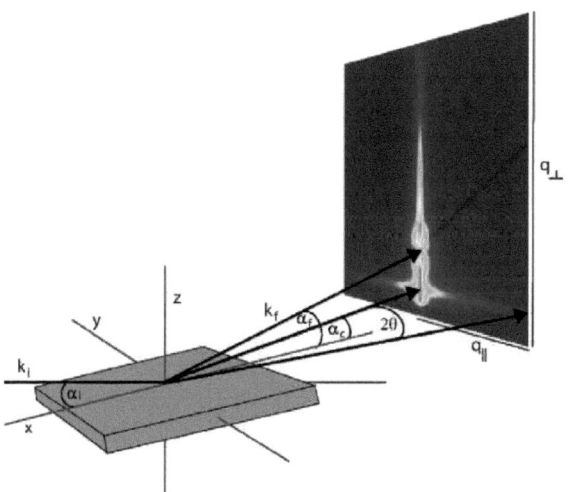

Figure II.13: GISAXS geometry

X-ray and neutron experiments
-
Introduction to grazing incidence small angle x-ray scattering (GISAXS)

In order to obtain lateral structural information, transverse detector scans along the reciprocal $q_\parallel$ (q_x, q_y) scattering plane at the Yoneda peak position ($\alpha_f = \alpha_c$) can be performed. Occurring peaks on the transverse $q_\parallel$ scans are often analyzed crudely with a generalized Bragg interference function $\xi_\parallel = \frac{2\pi}{q_\parallel}$, which is widely used for Born approximated scattering. However, for a rigorous data analysis of GISAXS one should use the DWBA. Theories for the most important film systems containing rough surfaces[75, 78, 79], buried particles[89] and supported islands[90] were previously derived. To compute and simulate GISAXS data, Lazzari developed the simulation and fitting software IsGISAXS[ii],[91, 92]. In this software the mentioned theories and approximations are included.

II.2.5.1 GISAXS geometry

In a typical GISAXS experiment the primary beam is directed in a certain incident angle α_i onto the sample's surface. The incident beam is reflected and scattered at the samples surface, with an exit angle α_f perpendicular to the surface and an angle θ parallel to the surface Figure II.13. The scattering wave vector q is composed of its single components q_x, q_y and q_z, related to the experimental angles by

$$q = \begin{pmatrix} q_x \\ q_y \\ q_z \end{pmatrix} = \frac{2\pi}{\lambda} \begin{pmatrix} \cos(\alpha_f)\cos(2\theta) - \cos(\alpha_i) \\ \cos(\alpha_f)\sin(2\theta) \\ \sin(\alpha_f) + \sin(\alpha_i) \end{pmatrix} \qquad (\text{II.2.22})$$

When absorption in the samples medium cannot be neglected, q_z becomes complex and is related to the imaginary part, β, of the refractive index $n = 1 - \delta + i\beta$ and the materials critical angle $\alpha_c = \arcsin(\sqrt{2\delta}) \approx \sqrt{2\delta}$ by[93, 94]

[ii] www.insp.jussieu.fr/axe2/Oxydes/IsGISAXS/isgisaxs.htm

X-ray and neutron experiments
-
Introduction to grazing incidence small angle x-ray scattering (GISAXS)

$$q_z = \tilde{q}_z = \frac{2\pi}{\lambda}\left(\sqrt{\sin^2\alpha_i - \sin^2\alpha_c + 2i\beta} + \sqrt{\sin^2\alpha_f - \sin^2\alpha_c + 2i\beta}\right) \qquad (\text{II.2.23})$$

Throughout the rest of this work the $q_\perp = \sqrt{q_x^2 + q_z^2}$ and $q_\parallel = \sqrt{q_x^2 + q_y^2}$ reciprocal planes are used to describe scattering, reflection and refraction perpendicular and parallel to the specimen's surface. In some publications $q_\parallel$ and $q_\perp$ are approximated with q_y and q_z, respectively. This approximation is usually allowed, because $q_x \ll q_y$ for small α_i and α_f.

II.2.5.2 The Distorted Wave Born Approximation

In order to reduce GISAXS analysis to lateral density fluctuations, detector scans in $q_\parallel$ are commonly performed at $\alpha_f = \alpha_c$ or $\alpha_i = \alpha_c$[86]. Analyzing transverse detector scans DWBA, which describes the four major scattering terms has to be considered (curly braces in Eq. (II.2.24); Figure II.14)[89].

$$\langle|\Psi^2|\rangle = \left\{\left|\Psi^{(0)} + \langle\Psi_s\rangle + \langle\Psi_d\rangle\right|^2\right\} + \left\{\langle|\Psi_s|^2\rangle - |\langle\Psi_s\rangle|^2\right\} + \left\{\langle|\Psi_d|^2\rangle - |\langle\Psi_d\rangle|^2\right\} + \left\{2\operatorname{Re}\left(\langle\Psi_s\Psi_d^*\rangle - \langle\Psi_s\rangle\langle\Psi_d^*\rangle\right)\right\} \qquad (\text{II.2.24})$$

Term 1: $|\Psi^{(0)} + \langle\Psi_s\rangle + \langle\Psi_d\rangle|^2$ Term 2: $\langle|\Psi_s|^2\rangle - |\langle\Psi_s\rangle|^2$ Term 3: $\langle|\Psi_d|^2\rangle - |\langle\Psi_d\rangle|^2$ Term 4: $2\operatorname{Re}(\langle\Psi_s\Psi_d^*\rangle - \langle\Psi_s\rangle\langle\Psi_d^*\rangle)$

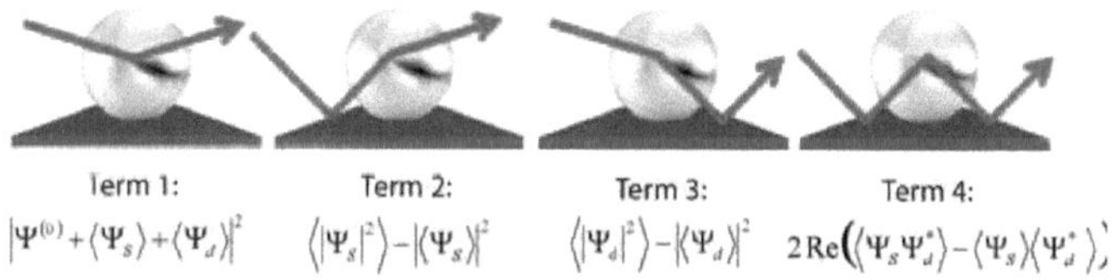

Figure II.14: The four possible scattering and reflection effects in the DWBA

$\Psi^{(0)}$ denotes the specular reflected amplitude from a smooth film surface without interior density fluctuations. The subscript S denotes the surface, while d denotes density fluctuations of the film's interior. $|\Psi^{(0)} + \langle\Psi_s\rangle + \langle\Psi_d\rangle|^2$ describes the scattered intensity at the specular beam position,

X-ray and neutron experiments
-
Introduction to grazing incidence small angle x-ray scattering (GISAXS)

$\langle |\Psi_s|^2 \rangle - |\langle \Psi_s \rangle|^2$ describes diffuse scattering from the surface roughness, $\langle |\Psi_d|^2 \rangle - |\langle \Psi_d \rangle|^2$ can be correlated to diffuse scattering from density fluctuations within the medium. The last term $2\,\mathrm{Re}\left(\langle \Psi_s \Psi_d^* \rangle - \langle \Psi_s \rangle \langle \Psi_d^* \rangle \right)$ includes possible correlations between scattered waves from surface roughness and density fluctuations and can in many cases be approximated to 0.

To relate scattered intensities with shapes of scattering objects, a relationship between a structure factor $\Gamma(q)$, which describes all the structural features in the observing length scales, and the scattered differential cross-section is needed. Sinha et.al.[75] derived the differential cross-section of diffuse scattering as

$$\frac{d\sigma}{d\Omega} = \frac{A k_c^2}{(4\pi)^2} \left| T^i T^f \right|^2 \Gamma(q) \qquad (\text{II.2.25})$$

where A is the irradiated sample surface area, k_c^2 is the critical impulse and T^i, T^f the Fresnel transmission coefficient of the incoming and outcoming wave, respectively, given by

$$T = \frac{2 k_\perp}{k_\perp + \tilde{k}_\perp} \qquad (\text{II.2.26})$$

Here $k_\perp$ and $\tilde{k}_\perp$ are the wave vectors perpendicular to the sample surface in vacuum and in the film medium, respectively. $\Gamma(q)$ can be related to density autocorrelation functions, which describe the form of idealized scattering objects. To deconvolute $\Gamma(q)$ Rauscher et.al.[89] used an infinitely thin and perfectly flat δ-layer with interior density fluctuations of a lateral distance $R_\parallel$ and obtained

$$\Gamma(q) = \int d^2 R_\parallel \exp(-i q_\parallel \cdot R_\parallel) C_w(R_\parallel) t^2 = \hat{C}_w(q_\parallel) t^2 \qquad (\text{II.2.27})$$

where $\hat{C}_w(q_\parallel)$ is the Fourier Transform of the density autocorrelation function $C_w(R_\parallel)$ of the lateral surface structures within this δ-layer, and t is the thickness of the δ-layer. In the case of non correlated density fluctuations, $\hat{C}_w(q_\parallel)$ can be substituted with idealized particle form factors[91]. Using the DWBA in combination with common form factors, GISAXS can be simulated can be

X-ray and neutron experiments
-
Introduction to grazing incidence small angle x-ray scattering (GISAXS)

simulated with e.g. the free available IsGISAXS software[91, 95] and compared with experimental results.

II.2.6 µ-x-ray reflectivity and µ-GISAXS on MC arrays at BW4

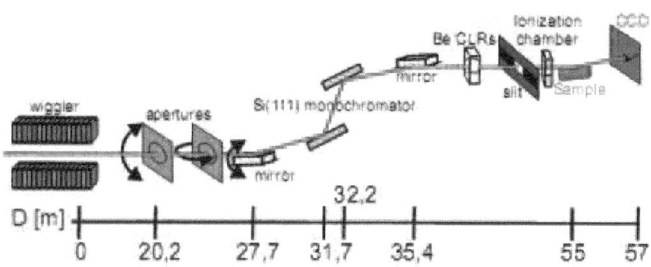

Figure II.15: Layout of the beamline BW4 at HASYLAB, DESY

Comparative to MC bending experiments, µ-GISAXS and µ-x-ray reflectivity studies on MC arrays were conducted at beamline BW4, HASYLAB at DESY, working at a wavelength of λ = 0.138 nm (Figure II.15).
The MC arrays were mounted into an environmental sample cell, which was developed within the scope of this thesis. The capton sealed cell could be purged with continuous streams of various gases, such as N_2, and heated to T = 150°C. Thus studies on structural changes in functional films upon temperature induced phase transitions became accessible. The sample cell was mounted onto a two circle goniometer equipped at an x/y/z translation stage positioned in a sample to detector distance of ~ 2.0 m. To be able to address single MCs with the illuminating X-ray beam, the micro focus option available at BW4 was used [23, 84].

X-ray and neutron experiments
µ-x-ray reflectivity and µ-GISAXS on MC arrays at BW4

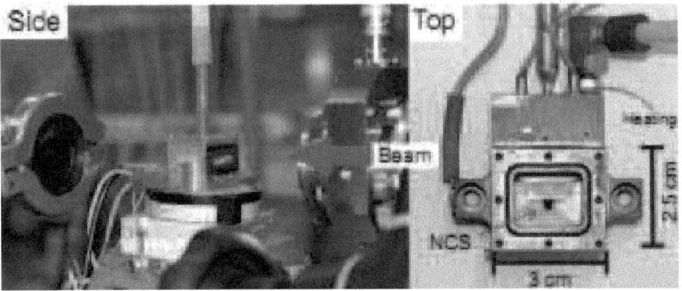

Figure II.16: Side and top view of the MC array placed in the made sample. The topview shows the dimensions of the sample cell made of brass, where the MC array is located. The heating of the cell could be controlled with a PT-100 temperature sensor and cooled with compressed air. The side view shows the sealed sample cell mounted on the goniometer. The x-ray beam can pass the two capton sealed windows. The sample cell can be purged with gases through the green tube which is attached to the cell's top cover.

The Gaussian FWHM beam dimension focused on the MC sensor was of a size of 32x17 µm² (horizontal x vertical), which matched the dimensions of a single MC. After alignment the MC array was tilted into the x-ray under a certain angle α_i (typical ~ 0.7°). Following the MC array was scanned in y-direction and the x-ray intensity on the primary beamstop at $\alpha_i + \alpha_f = 0°$ was measured. From such scans typical intensity vs. motor position profiles (Figure II.17a) were recorded. When the array was not placed in the x-ray beam, the beam's full intensity was transmitted to the beamstop with 2000 counts (graphs left hand side). Then the MC array was moved into the x-ray beam and the intensity dropped to 1000 counts. After further movement of the MC array into the x-ray beam the MC *"C1"* was hit by the x-ray beam and the rectangular structure of the MC bar was convoluted with the Gaussian beam profile, which led to the V-shaped intensity profiles for the single MC's.

Measurements at constant incident angles were performed in order to obtain typical GISAXS patterns (Figure II.17). Integration times were typically ~ 120 min. Transverse detector scans at the samples critical angle α_c, were performed in order to extract scattering in the $q_\parallel$ scattering plane (chapter II.2.5).

X-ray and neutron experiments

μ-x-ray reflectivity and μ-GISAXS on MC arrays at BW4

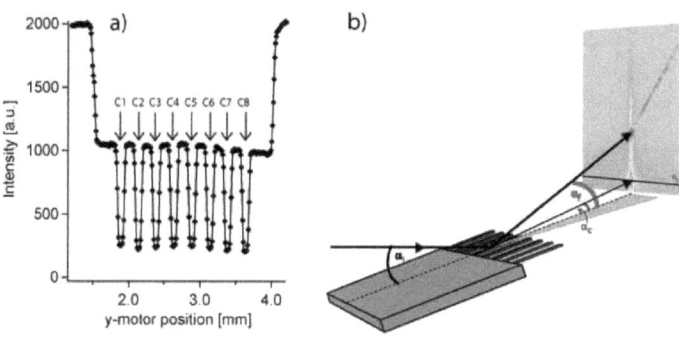

Figure II.17: a) Intensity plot in respect to y-motor position; the tilted MC array is scanned along its x-direction. The x-ray beams primary intensity is measured on the primary beamstop at $\alpha_i + \alpha_f = 0°$. Thus a decrease in transmitted intensity denotes the y-position of one single MC (C1-C8). Knowing the positions of all MCs allows to record GISAXS patterns for each individual MC; b) Scattering geometry of GISAXS on one single MC

In reflectivity scans the MC was illuminated for a short time at $0.1° \leq \alpha_i \leq 1.8°$ with intervals of $\Delta\alpha \approx 0.015°$. Thus more than 100 2-D detector images had to be analyzed in order to obtain one reflectivity profile. Hence, an automatic script based routine had to be used to analyze the obtained detector images and construct a reflectivity profile.

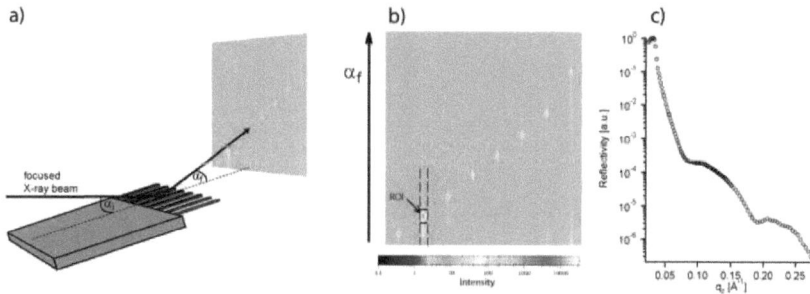

Figure II.18: a) Geometry of a focused x-ray beam reflected at a single MC; The reflected beam profile is detected at the 2-D CCD detector at $\alpha_f = \alpha_i$; b) Selected detector image parts at various α_f plotted in parallel; The ROI is defined according to the Gaussian profiles FWHM in the $q_\perp$ and $q_\parallel$ plane; c) typical reflectivity profile obtained after image processing and data treatment

X-ray and neutron experiments

µ-x-ray reflectivity and µ-GISAXS on MC arrays at BW4

Every obtained 2-D detector image was first normalized for the intensity per second of the incoming x-ray beam measured at the ionization chamber. This allowed normalizations in respect to signal integration times and the used absorbers, which were placed between the slit and the ionization chamber. Following the beam profile dimensions were analyzed along the $q_\parallel$ and $q_\perp$ plane with a Gaussian function and the boundaries for the region of interest (ROI) were set according to the profile's full width half maximum (FWHM). The pixels intensities in the ROI were integrated and one dimensional intensity vs. q_z data was obtained. The whole curve was normalized to the reflected intensity at $\alpha_i < \alpha_c$. As a result 1-D reflectivity data was obtained.

II.2.7 Neutron reflectivity at N-REX⁺ (FRM II)

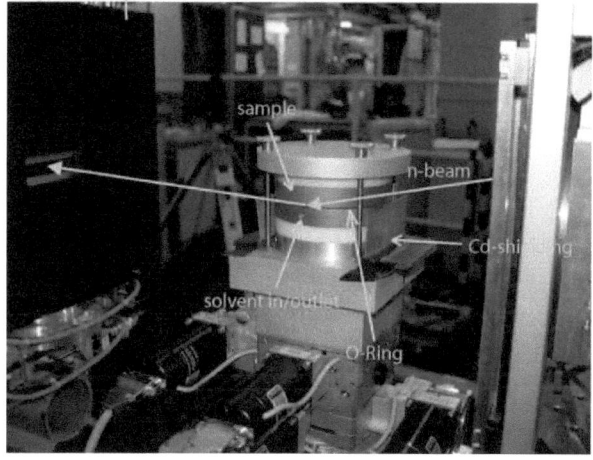

Figure II.19: Experimental setup of the N-REX⁺ reflectometer *(up)*; the monochromatic neutron beam passes a horizontal and vertical adjustable aperture and is reflected at the Si/polymer/solvent interfaces; neutrons are reflected towards the 2-D detector passing a second aperture and a Cadmium shielding tube; *down)* close-up of goniometer mounted liquid sample cell; the solvent is trapped between the sample and a second *Si* disk, which contains two drilled holes for liquid injection and exchange. The space between the sample and the *Si* disk is sealed with a 1 mm thick O-ring.

X-ray and neutron experiments

Neutron reflectivity at N-REX+ (FRM II)

Neutron reflectivity experiments on poly-methyl-methacrylate (PMMA) endgrafted polymer brushes were conducted comparative to surface stress investigations at the N-REX$^+$ reflectometer located at the Forschungs Reaktor München II (FRM II). FRM II operates at a thermal power of 20 MW. Fast neutrons are cooled down in the reactor core and in a moderator tank filled with D_2O. Cold (slow) neutrons are guided through neutron beam guidance tubes to a graphite monochromator. Diffracted neutrons are of a wavelength of λ = 4.26 Å. The beam passed a horizontal and vertical adjustable aperture, which was set to a gap size of 20 mm horizontal x 1 mm vertical. For intensity reductions of the reflected beam at small α_i the vertical aperture gap was decreased to 0.02 mm. Passing the aperture the neutron beam was guided to the polymer brush sample (aperture/sample distance: ~ 300 mm), which was mounted in a liquid cell.

The neutrons were transmitted through the Si substrate and reflected at the Si/PMMA-brush/solvent interfaces. For high density contrasts the solvent environment was fully deuterated, while the polymer was fully hydrogenated. The neutron flux at the sample place was estimated to be ~$3 \cdot 10^6$ n·cm^{-2}s^{-1}.

Reflectivity scans were performed in the range of $0° \leq \alpha_i \leq 1.6°$ with $\Delta\alpha_i = 0.02°$. For background reduction cadmium shielding was applied to the sample cell up- and downwards the neutron stream. Reflected neutrons passed a second aperture (sample/aperture distance: ~ 300 mm), which could be optimized for background reduction (gap size ~ 1 mm) or fully opened for the detection of off specular scattering. Further background reduction was achieved by the application of a cadmium shielded tube between the second aperture and the detector (sample to detector distance: 2464.5 mm).

X-ray and neutron experiments

Neutron reflectivity at N-REX+ (FRM II)

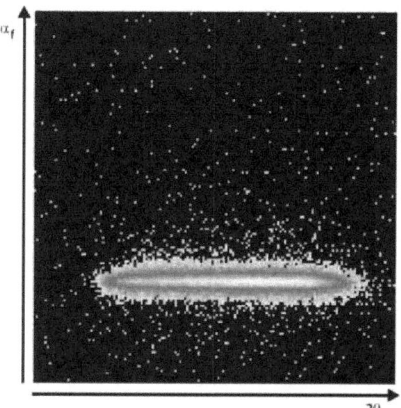

Figure II.20: Typical 2-D detector image for constant α_i.

In order to obtain common reflectivity profiles one 2-D detector image (Figure II.20) was recorded at each adjusted α_i. Thus, for a complete reflectivity scan a series of ~ 80 2-D detector images were recorded. Obtained detector images were further processed by integrating along 2θ. Thus one dimensional α_f dependent intensity data was obtained for each single α_i. After integration of all images belonging to one reflectivity scan three dimensional intensity data could be obtained in respect to α_i and α_f. Before plotting the data sets in a q_x/q_z contour plot, the reflected intensity was corrected for the beam's elongation, which exceeded sample dimensions at small α_i. Horizontal line scans crossing the first oscillatory maximum were performed in order to analyze the reflected beam profile. Horizontal scans at the maximum, which is located at $q_x = 0 \: Å^{-1}$, yield the specular reflected intensity depending on q_z. Off specular scans along q_z at ~ 5-10% of the maximum reflected intensity contain information on the background and possible interface correlations, as discussed in chapter II.2.4. For background corrections of specular reflected intensities and in order to be able to estimate absolute roughness values, specular reflected intensities were corrected for off specular intensities by subtraction.

X-ray and neutron experiments

Neutron reflectivity at N-REX+ (FRM II)

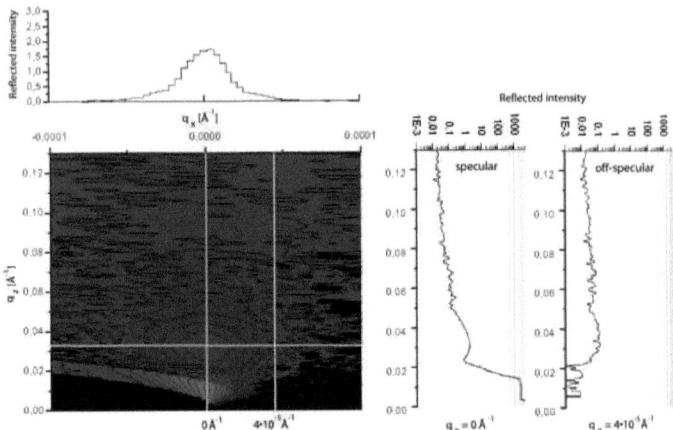

Figure II.21: typical q_x/q_z contour plot obtained from one reflectivity scan. The horizontal line crossing the first oscillatory maximum represents the reflected beam profile. Vertical line scans along q_z at $q_x = 0$ Å^{-1} yield the specular reflected intensity, while linescans performed at $q_x = 4 \cdot 10^{-5}$ Å^{-1} yield the offspecular reflected intensity.

II.3 Thermodynamics of mixing

II.3.1 Free energy of mixing

Within the framework of this thesis grafted polymer films were studied, which changed their properties upon mixing and demixing processes. In the first case an end grafted PMMA polymer brush system was studied which was found in a collapsed state in bad solvent environments and swollen in good solvent environments. In the second case, the demixing process of a surface grafted miscible poly-styrene (PS)/poly-vinyl-methylether (PVME) film upon temperature increase was studied. Brush swelling/collapse and polymer blend mixing/demixing processes are driven by the Helmhotz free energy of mixing, which is defined as

$$\Delta F_{mix} = \Delta U_{mix} - T\Delta S_{mix} \qquad (\text{II.3.1})$$

The total internal energy change of mixing for two species A and B can be derived within the regular solution theory as a function of three pairwise interaction energies (u_{AA}, u_{AB}, and u_{BB}):

$$\Delta U_{mix} = \frac{z}{2}\phi(1-\phi)(2u_{AB} - u_{AA} - u_{BB}) \qquad (\text{II.3.2})$$

where z is the number of nearest neighbors in a particle lattice (Figure II.22) and ϕ is the volume fraction of species A. Using Eq. (II.3.2) the dimensionless Flory interaction parameter χ can be defined to characterize the difference of interaction energies in the mixture:

$$\chi \equiv \frac{z}{2}\frac{(2u_{AB} - u_{AA} - u_{BB})}{kT} \qquad (\text{II.3.3})$$

The kT normalized χ-parameter is a measure to determine, if the net interaction between two species is attractive or repulsive.

When $2u_{AB} < u_{AA} + u_{BB}$, the χ-parameter is smaller 0, there is net attraction between two lattice sites occupied by two different species and a single-phase mixture is favored. In the other case when $2u_{AB} > u_{AA} + u_{BB}$, the χ-parameter is bigger 0, there is net repulsion between two lattice sites

Thermodynamics of mixing
-
Free energy of mixing

occupied by two different species and a two-phase mixture is favored. However, such conclusions are only valid, when the entropic part $T\Delta S_{mix}$ is negligibly small compared to the interaction part.

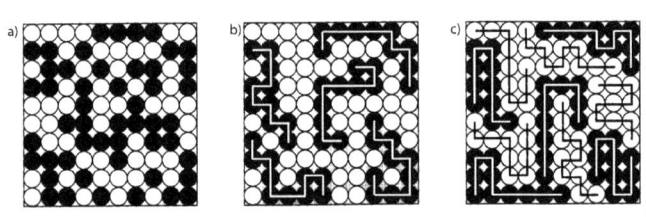

Figure II.22: Illustrative scheme of the particle lattice model used for derivations for free energies of mixing; a) regular solution of 50 (solvent) particles of type A and 50 (solvent) particles of type B. There are no covalent connections between the particles. Degrees of freedom and resulting entropy of mixing is maximized; b) polymer solution of chains composed of 10 covalently linked monomer units; degrees of freedom and entropy of mixing decreases; c) polymer blend of 5 chains of polymer type A and 5 chains of polymer type B; degrees of freedom are entropy of mixing is minimized[iii].

The entropy of mixing can be derived as

$$\Delta S_{mix} = -k\left[\frac{\phi}{N_A}\ln\phi + \frac{(1-\phi)}{N_B}\ln(1-\phi)\right]$$ (II.3.4)

where N_A and N_B are the numbers of lattice sites occupied by molecule A and B. Since $\phi < 1$, the entropy of mixing is always bigger 0. For regular solutions of low molecular weight molecules, such as two different solvents $N_A = N_B = 1$, the degrees of freedom and resulting entropies of mixing are maximized (Figure II.22 a). For polymer solutions $N_A = N >> 1$ and $N_B = 1$, with N the number of monomers per polymer chain. One can see that for $N_A \rightarrow \infty$ and $N_B = 1$, ΔS_{mix} reduces to:

$$\Delta S_{mix} = -k[(1-\phi)\ln(1-\phi)]$$ (II.3.5)

Accordingly the degrees of freedom reduce and the entropy of mixing decreases (Figure II.22). Hence, for polymer blends with $N_A \rightarrow \infty$ and $N_B \rightarrow \infty$, $T\Delta S_{mix}$ can be approximated as small

---

[iii] adapted from **Rubinstein, M.; Colby, R. H.,** *Polymer Physics.* Oxford Univ. Press: Oxford, 2003; p XI, 440 S.

Thermodynamics of mixing
-
Free energy of mixing

compared to ΔU_{mix} (Figure II.22 c).

II.3.2 Collapsed/stretched polymer brushes

In the last 20 years well derived and widely accepted theoretical models for polymer segment distributions, scaling laws and resulting free energies for polymer brushes appeared in literature[26, 27, 30, 68, 96-98]. Such models were able to explain the collapsed and stretched polymer brushes, according to the environmental solvent conditions in detail. Within this chapter the theoretical background is discussed, which is needed to explain the obtained comparative reflectivity and surface stress data obtained within thesis. Using the discussed theoretic approaches in combination with the obtained experimental data it becomes possible to gain valuable physical understanding of the mechanics in polymer brush films (chapter V).

II.3.2.1 Polymer brush: Definition and models

Polymer brushes are polymer systems, where polymer chains are tightly bound at one end to a surface. Polymer chains can be bound to e.g. curved colloidal particles or flat substrates, such as pieces from Si-wafers. Among the variety of possible polymer brush systems, this thesis deals with poly-methyl-methacrylate (PMMA) polymer brushes, which are chemically bound at one single end to a flat Si-surfaces. The polymer chains were "grafted from" the substrate. This means that the PMMA polymerization started from a surface bound polymerization initiator. The surface initiated polymerization was performed with Atomic Transfer Radical Polymerization (ATRP), which is discussed in detail in chapter V.2.1. The chosen polymerization route allowed creating polymer brushes of small polydispersities, which are laterally densely grafted.

In fact the density of grafting is one of the most important parameters characterizing a brush system, because highly grafted brushes tend to be stretched away from the surface in good solvent environment. Such behavior results in polymer brush heights H which are bigger than the polymer's radius of gyration R_g.

Thermodynamics of mixing - Collapsed/stretched polymer brushes

There are some common definitions defining the grafting density of a brush. In this work the grafting density is defined as the area A occupied by one grafted polymer chain in respect to the cross sectional area of one chain, which can be expressed by the square of the statistical segment length l:

$$\sigma = A/l^2 \qquad (\text{II.3.6})$$

The grafting density can be directly calculated from x-ray reflectivity experiments of the dry polymer brush, by fitting a step-profile function to the reflectivity profile. From the real part of the refractive index (Eq. (II.2.10)) and the brush height H one can calculate the mass density per unit area ρ^A. σ can then be expressed as:

$$\sigma = \frac{M_w}{H\rho} \cdot \frac{1}{N_A l^2} = \frac{M_w}{\rho^A} \cdot \frac{1}{N_A l^2} \qquad (\text{II.3.7})$$

where M_w is the molecular weight of the polymer.

The stretching behavior of densely grafted polymer brushes in good solvents was first explained by Alexander and deGennes[26, 99]. The Alexander-deGennes model describes this phenomenon with the equilibration of two losses in free energy. On the one hand the polymer chain loses conformational entropy by the stretching process. On the other hand, chain entanglements reduce the energetic favored interactions between polymer segments and solvents molecules. The total free energy is therefore a sum of an elastic energy (F_{el}) term and an internal energy (F_{int}) term:

$$F = F_{el} + F_{int} \qquad (\text{II.3.8})$$

However, the Alexander de-Gennes model makes two simplifying approximations. First, a step-profile function is assumed, which leads to constant segment densities ϕ in the polymer brush. Second, the polymer chains are assumed to be stretched equally, which leads to brush solvent interfaces with statistical roughness profiles described by *erf* or *tanh* functions (Eqn. (II.2.15), (II.2.16)).

Thermodynamics of mixing
-
Collapsed/stretched polymer brushes

On the basis of the Alexander de-Gennes (AdG) model, Milner, Witten and Cates (MWC)[68, 97, 98] and Zhulina et al.[69] described the stretching behavior of polymer brushes with a parabolic type potential. The main difference to the AdG model is that the chain ends are allowed occupying arbitrary positions in the polymer brush phase. This leads to non constant segment distributions $\phi(z)$ with parabolic roughness profiles.

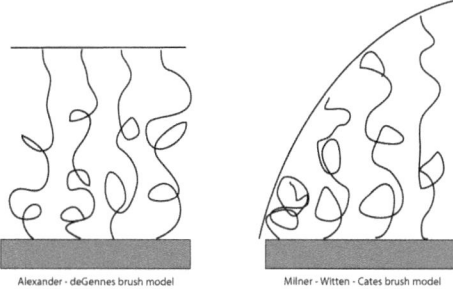

Figure II.23: Comparison of the step type Alexander – deGennes model (left) with the Milner-Witten-Cates parabolic type brush model (right)

It turned out that both the brush height H and free energy F scale equally with the degree of polymerization N, according to

$$H_{MWC} \propto H_{AdG} \propto N \qquad (\text{II.3.9})$$

$$F_{MWC} \propto F_{AdG} \propto N \qquad (\text{II.3.10})$$

Comparing prefactors from AdG scaling laws with MWC scaling laws following relationships are obtained[98]

$$\frac{H_{MWC}}{H_{AdG}} = 1.3 \quad \text{and} \quad \frac{F_{MWC}}{F_{AdG}} = 0.89 \qquad (\text{II.3.11})$$

Segment profiles in bad solvent conditions are better described with a step type profile[96].

Thermodynamics of mixing
Collapsed/stretched polymer brushes

II.3.2.2 Collapse-stretching of polymer brushes in mixed solvents

For dense grafted polymer brushes Birshtein and Lyatskaya[30] developed a model to calculate free energies and resulting brush heights for mixtures of good and poor solvent conditions. In such a way the description of the collapse/stretching mechanism became possible within one theoretical approach. They used standard equations from Flory-Huggins to describe the free energy of the brush in the arbitrary solvent mixtures.

A thermodynamic equilibrium condition of the solvent molecules A and B in the brush and the bulk is presumed:

$$\mu_x^{brush} = \mu_x^{bulk} \qquad x = A, B \qquad (\text{II.3.12})$$

The volume fractions of the single components in the brush Eq. (II.3.13) and in the bulk Eq. (II.3.14) are related to

$$\phi = 1 - \phi_A - \phi_B \qquad (\text{II.3.13})$$
$$\phi_A^{bulk} + \phi_B^{bulk} = 1 \qquad (\text{II.3.14})$$

The terms for the free energies can be expressed as functions of the binary interaction parameters χ_A, χ_B, which describe the pair wise interaction of the polymer with good solvent A and bad solvent B, and χ_{AB}, which describes the solvent-solvent interactions. The sums of the internal and elastic free energies in the brush and bulk were derived as:

$$F^{brush} = n_A \ln \phi_A + n_B \ln \phi_B + \phi(\chi_A n_A + \chi_B n_B) \\ + n_A \phi_B \chi_{AB} + 3H^2/N \qquad (\text{II.3.15})$$

$$F^{bulk} = n_A \ln \phi_A^{bulk} + n_B \ln(1 - \phi_A^{bulk}) + n_A(1 - \phi_A^{bulk})\chi_{AB} \qquad (\text{II.3.16})$$

With the equilibrium condition from Eq. (II.3.12) and $\dfrac{\partial F^{brush,bulk}}{\partial n_x} = \mu_x^{brush,bulk}$ a binary equation system can be obtained, which is only dependent on the solvent composition in the bulk, the Flory interaction parameters χ_A, χ_B and χ_{AB} and the grafting density σ:

$$\ln\left(\phi_A/\phi_A^{bulk}\right) + \chi_{AB}\left[\phi_B(1-\phi_A) - \left(1-\phi_A^{bulk}\right)^2\right]$$
$$+\phi + \left[\chi_A(1-\phi_A) - \chi_B\phi_B\right]\phi + 3/\phi\sigma^2 = 0 \qquad (\text{II.3.17})$$

$$\ln\left(\phi_B/\left(1-\phi_A^{bulk}\right)\right) + \chi_{AB}\left[\phi_A(1-\phi_B) - \phi^2\right]$$
$$+\phi - \left[\chi_A\phi_A - \chi_B(1-\phi_B)\right]\phi + 3/\phi\sigma^2 = 0 \qquad (\text{II.3.18})$$

Defining an effective polymer-solvent interaction parameter χ_{eff}:

$$\chi_{eff} = \chi_A \phi_A^{bulk} + \chi_B\left(1-\phi_A^{bulk}\right) - \chi_{AB}\phi_A^{bulk}\left(1-\phi_A^{bulk}\right) \qquad (\text{II.3.19})$$

Comparative surface stress - ϕ_A^{bulk} data can be transformed in surface stress data depending on effective polymer-solvent interaction parameters. By such procedure valuable insights in brush mechanics on bad ($\chi > 0.5$), θ ($\chi = 0.5$) and good ($\chi < 0.5$) solvent conditions could be drawn (chapter V.4).

II.3.3 Phase separating polymer blends

Polymer films, which undergo phase transitions with changing environmental conditions, such as temperature, pH, and solvent quality changes are promising systems for switching surfaces[100]. Besides the collapse/stretching behavior of endgrafted polymer brushes on the solvent environment, phase transitions in grafted polymer films caused by temperature changes were studied. In literature one can find the surface grafted poly-n-isopropylacrylamide (PNIPAM) brush, as an example for a system, which undergoes Lower Critical Solution Temperature (LCST) behavior, when immersed in water[101-103]. However, densed grafted polymer brushes showed similar LCST behavior than in bulk[102]. In this work an alternative route for the preparation of surface grafted polymer films is

Thermodynamics of mixing
-
Phase separating polymer blends

searched, where the LCST behavior of a grafted demixing polymer blend can be tuned by the density of grafting points.

The first task for such an approach is to search for a suitable polymer blend film system, which shall be grafted in later stages. There are several different types of polymer blends, which mix or demix with increasing temperature. They are classified by the temperature dependence of the polymer-polymer interaction parameter $\chi(T)$. For idealized symmetrical phase diagrams, $\chi(T)$ can be empirically expressed as a linear function of $1/T$ with a temperature independent term A and a temperature dependent term B/T:

$$\chi(T) = A + \frac{B}{T} \qquad (\text{II}.3.20)$$

Within this linearity condition two different kinds of mixing/demixing polymer pairs exist. The first type has negative values for the parameter A and positive values for parameter B. Such values lead to a decrease of the χ-parameter with increasing temperatures. Consequently the polymer blend turns from a two-phase system into a mixed one phase region by rising temperatures. Such phase transition behavior is called Upper Critical Solution Temperature (UCST) behavior (Figure II.24 a). One example for thin polymer films exhibiting UCST behavior above room temperature (RT) is the deuterated polystyrene (dPS)/poly(p-methylstyrene) (PpMS) blend[104]. However, for the purpose of surface grafting a polymer film system was searched, which forms homogeneous films. There are only few polymer pairs, which form a one phase films at room temperature and dewet at higher temperatures, which is called Lower Critical Solution Temperature (LCST) behavior (Figure II.24b). Temperature dependent B parameters have to be highly negative for polymer blends, which form uniform films at RT and phase separate at temperatures well above RT. One such example is the dPS/poly-vinyl-methyl-ether (PVME) polymer blend. Here parameter values of $A = 0.0973$ and $B = -41.6$ K can be found[105]. According to Eq. (II.3.20) the χ-parameter is negative at T < 155°C[105] (Figure II.25).

Thermodynamics of mixing

Phase separating polymer blends

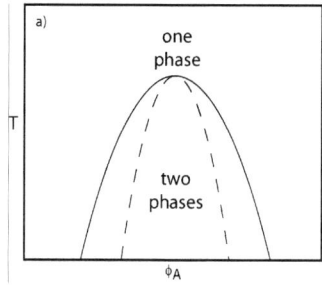
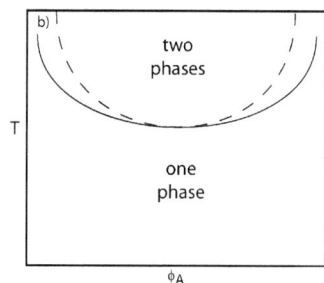

Figure II.24: a) scheme of a symmetric phase diagram of a polymer blend with UCST behavior; b) symmetric phase diagram of a polymer blend with LCST behavior; straight lines are binodals, separating the one phase region from the metastable region; dashed lines represent the spinodal lines, which separate the metastable from the two-phase region; binodal and spinodal lines coincide at a critical point at $\phi_A = 0.5$, where no metastable phase is present.

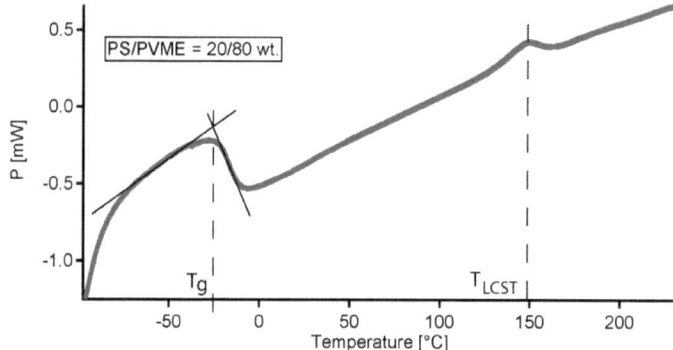

Figure II.25. Differential Scanning Calorimetry (DSC) thermogram of a thick PS/PVME blend film cast from toluene; films composed of PS/PVME = 20/80 wt. were found to be fluid at room temperature ($T_g <$ 0°C), with a $T_{LCST} < 150$°C; (M(PS) = 26700 g/mol; M(PVME) = 66000 g/mol)

One more advantage for the purpose of homogeneous grafting of the PS/PVME blend is the single glass transition temperature, T_g, of bulk films cast from certain solvents, such as toluene[19]. Since the PS homopolymer has a $T_g = 100$°C and the PVME homopolymer has a $T_g = -40$°C, the polymer blends T_g is steadily decreasing with increasing PVME volume fraction. In such a way T_g values below 0°C can be found for the PS/PVME = 20/80 wt. blend (Figure II.25).

Thermodynamics of mixing
-
Phase separating polymer blends

However, experimentally found phase diagrams of PS/PVME bulk blend films were found to be asymmetric and dependent on the molecular weight of the polymeric components. Nishi and Kwei[21] observed that the cloud-point indicating a phase separation temperature is decreasing with increasing molecular weights. Such behavior can be explained with reduced entropies of mixing for higher molecular weight polymers. However, for molecular masses > 100000 g/mol the cloud point did not decrease essentially. Such behavior was addressed to entanglement effects, which counteract two phase formations.

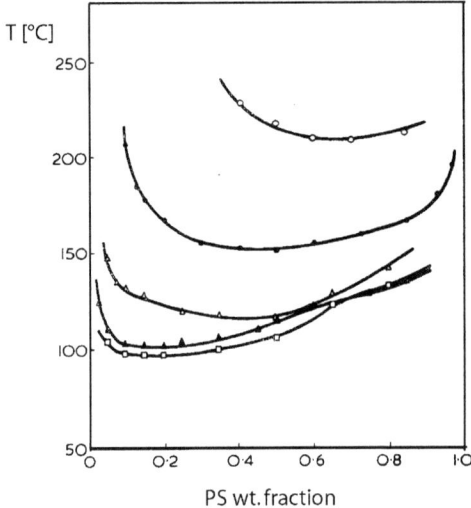

Figure II.26: Cloud point curves for PS/PVME thick film blends with changing weight average molecular weight M_w(PS) empty circles: 10000; filled circles: 20400; empty triangles: 51000; filled triangles: 110000; empty boxes: 200000 (taken from Nishi and Kwei[21])

For the application of the PS/PVME blend film in grafted film applications, which have thicknesses of only one molecular layer, the thickness dependence on the thermodynamic blend properties have to be investigated. Especially in films of thicknesses smaller the polymers R_g, it turned out that wall confinements lead to increasing coil densities and reduced entanglements[106-109]. Such chain confinements may not only lead to a decrease in T_g[110-113], but influence also the polymer blends LCST. The first study on film thickness dependent polymer blend LCST behavior was first studied by Reich and Cohen[114]. They used experimental cloud point studies on PS/PVME films deposited

Thermodynamics of mixing
-
Phase separating polymer blends

on charged glass and uncharged gold substrates in order to explain qualitatively, film thickness dependent phase separation behavior towards three major effects:

i) Geometric effects, resulting from high film surface to film bulk ratios lead to increasing cloudpoint temperatures with decreasing film thicknesses.

ii) Selective adsorption of PS to the substrate leads to decreasing cloudpoint temperatures with decreasing film thicknesses for PS concentrations higher than the bulk's critical concentration (concentration with the lowest cloud point temperature in the 3-D bulk phase diagram).

iii) Charged substrates have certain surface potentials, which decay exponentially with film thickness. Such fields were found to destabilize the PS/PVME mixture, and led to decreasing cloud point temperatures with decreasing film thicknesses.

On the basis of Reich and Cohen's work, Tanaka et al.[115] experimentally investigated blend compatibilities in ultra-thin two dimensional PS/PVME films of thicknesses smaller twice the bulk's R_g of the longest component. They observed the formation of two phase systems already below room temperature for PS/PVME films in ultra-thin thickness regions. Verifying the preferential adsorption of PS to the SiO_x-surface with secondary ion mass spectroscopy (SIMS) it was possible to explain such behavior with a change in chain conformation from the 3-D film towards the 2-D film. The PVME, which was enriched at the film/air interface, was in a conformationally constrained state. To overcome such entropically unfavoured conformational states, the PVME phase separated from the PS at temperatures below RT and thus by minimum 50°C lower than for the thinnest prepared 3-D film. Within this phase separation process PVME chains retrieved partly their energetically favored random coil state (Figure II.27). However, experiments conducted within this work show that such phase separation within the 2-D ultra thin PS/PVME film can be shifted to $T > RT$, when the polymer chains are irreversibly grafted to the substrate. Employing grafting points to the polymer chains reduce the PVME's ability to recover its conformational energy, which leads to higher LCSTs. Moreover, obtained results predict that the LCST can be possibly tuned by the amount of grafting points per polymer chain.

Thermodynamics of mixing
-
Phase separating polymer blends

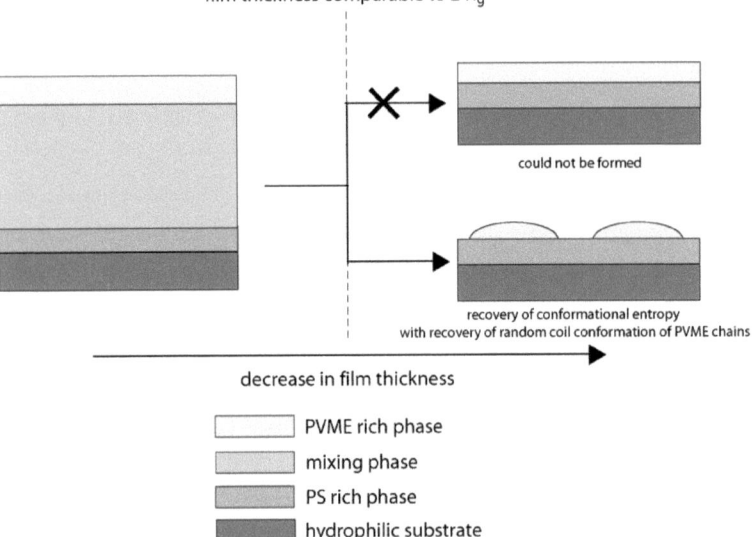

Figure II.27: Schematic representation for the formation of the phase separated microstructure for 2-D PS/PVME films. In 3-D thin films of $t > 2R_g$ the PS is enriched at the substrate/film interface, followed by a mixed phase and the PVME enriched at the film/air interface. Reaching the 2-D ultra-thin thickness regime a bilayered PS/PVME structure cannot be formed. In contrast PVME enriched droplet formation by conformational random coil recovering is observed[iv].

[iv] adapted from Tanaka, K.; Yoon, J. S.; Takahara, A.; Kajiyama, T., Ultrathinning-Induced Surface Phase-Separation of Polystyrene Poly(Vinyl Methyl-Ether) Blend Film. *Macromolecules* **1995**, 28, (4), 934-938

II.4 Miscellaneous experimental techniques

II.4.1 Contact angle experiments

Contact angle experiments were performed on various specimen using a Krüss, DSA10-MK2. The static contact angle was measured in air atmosphere at room temperature with various solvents.

II.4.2 Gel Permeation Chromatography

Gel permeation chromatography (GPC) was used to determine the molecular weight and polydispersity index of synthesized and purchased polymers. GPC is a chromatographic technique, which fractionates the polymer. The polymer is fractionated by diffusion into pores of different sizes, which is dependent on the molecular weight. Accordingly different retention times are obtained. A prior calibration is needed to assign the molecular weight. For this thesis the following devices were used: PSS (SDV) PS column, Waters 590 pump; as detecting unit a RI ERMA Inc. ERC 7512 ERC detector and UV S3702 (254 nm) detector were used. The eluent was toluene in the case of polystyrene and polyvinylmethylether and THF in the case of PMMA. The flow rate was set to 1 mL/min.

II.4.3 Differential Scanning Calorimetry

Differential scanning calorimetry (DSC) was performed to determine thermodynamical properties of bulk material. A thermally stable reference is heated in order to obtain a constant temperature ramp. The sample's temperature is measured in respect to the reference's temperature. The sample is heated in order to match the temperature of the reference. The difference in heating power, which is needed to obtain equal sample and reference temperatures, is monitored. Certain peaks or steps in the obtained thermograms yield information on exothermic and endothermic phase transitions in the sample material. DSC measurements were performed using a Mettler Toledo DSC822 under constant N_2 streams of 30 mL/min. Heating and cooling rates of 2°C/min and 10°C/min were used for measuring thermograms of PS/PVME and benzophenone linker bulk systems.

II.4.4 Environmental scanning probe microscopy (SPM)

Scanning probe microscopy (SPM) can be used to scan the surface topography with a small tip either in contact or intermitted mode (tapping mode). Images can be analyzed for surface roughness. Extracted height profiles yield information on domain sizes, forms and distances.

Environmental SPM of grafted PS/PVME blend were recorded in tapping mode at RT and at 150°C in a defined environment of N_2 gas (environmental SPM, Veeco Instruments, Santa Barbara, CA, USA). For these studies silicon cantilevers were used having a nominal spring constant of 42 N/m and a resonance frequency of around 300 kHz (Olympus, OMCL-AC160TS, Japan). For operation of the SPM a NanoScope IIIa controller (Veeco Instruments, Santa Barbara, CA, USA) controlled by the software 5.30r2 was used. The offset and tilt background of all images were removed by processing all images with a first order flattening procedure.

II.4.5 White light confocal microscopy

In contrast to bright field (conventional) microscopy, confocal microscopy allows to observe a defined spot on the specimen due to point illumination and blocking of out-of-focus information. Thus, and in contrast to conventional microscopy, the contrast and the depth resolution is increased up to a few nm. Similar to SPM, three-dimensional topographies can be reconstructed with less resolution, but from bigger areas.

A µSurf© white-light confocal profilometer (Nanofocus AG, Germany) was used to image the surface topography. The light source is an external xenon lamp which illuminates the sample through the microscope. An objective with 100× magnification was used. The corresponding maximum measurement area is 160×154 mm^2 and the vertical resolution is 1.5 nm.

II.4.6 X-ray reflectivity from lab x-ray sources

X-ray reflectivity experiments on coated Si-wafer samples were performed at a θ-θ XRD 3003 (Seifert Ltd., GB) diffractometer. Monochromatic and collimated Cu-Kα (l = 1.54Å) x-rays were supplied by a Cu-Anode. For heating experiments a custom-made oven was applied to the

Miscellaneous experimental techniques

experimental setup and measurements were performed at RT and 150°C under a constant flow of N_2 until $2\theta_{max} = 6°$. The applied crystal optics in combination with the used apertures allowed to record reflectivity smaller $1\cdot10^{-7}$.

II.5 Substrate preparation

II.5.1 Substrate cleaning

For cleaning porposes and in order to provide defined SiO_x surfaces for all studied samples, the used Si substrates (MC sensor arrays and wafers) were cleaned using the following protocol.
In a pre-cleaning step Si wafer substrates were sonicated in CH_2Cl_2 solution for 15 min. Fragile MC sensor arrays were cleaned in a stirred solution of CH_2Cl_2 for 15 min.
A mixture of 1:1:12.5 of conc. NH_3:H_2O_2:MilliQ water was preheated to 80°C. Following the substrates were added and removed after 25 min of constant temperature. While further processing the cleaned substrates were kept under MilliQ water for conservation of active SiOH surface sites (typically less than 3 hours).

II.5.2 Preparation of passivating Au films

For selective coating of MC sensor arrays topsides, passivating Au films were evaporated at the MC sensor arrays backside. For the prepared PMMA brushes on MC sensor arrays, half of the arrays topside was protected by an evaporated Au film.
The MC sensor arrays backside was coated with a 20 - 30 nm thick protecting gold film by thermal evaporation at a constant rate of 0.1nm/s and $p \sim 1.8 \cdot 10^{-5}$ mbar (BALTEC MED 020, BALTEC, Balzers, Lichtenstein).
For topside protection, half of the MC arrays topside was coated as well with 20 - 30 nm protecting gold films with the use of a shadowing mask.

II.6 Materials

- Methylmethacylate (MMA) (Acros, 99%) was purified by passing through an alumina column and distilled under reduced pressure and stored under argon at -20°C.
- Anisole (Aldrich, 99%) was saturated with Ar by passing a continuous Ar stream through the liquid.
- CuBr (Aldrich, 98%) was purified by boiling in mixture of 1:1 (by volume) Millipore water/acetic acid and subsequently filtered off. The precipitate was rinsed with water, ethanol, and finally with diethyl ether and dried in a vacuum oven for 24 h.
- N, N, N', N', N''-pentamethyldiethylenetrieamine (PMEDTA) (Aldrich,99%) was purified by destillation under reduced pressure.
- Ethyl 2-bromoisobutyrate (2-EiBBr) (Aldrich, 98%) was used without further cleaning.
- Triethylamine (NEt$_3$) was distilled and stored under Argon atmosphere.
- The ATRP starter (4), 3-(2-bromoisobutyryl)propyl)dimethyl-chloro-silane was synthesized following the procedure described in literature[116] using reagents as received and purified by distillation under reduced pressure. In order to prevent its degradation by moisture, the ATRP starter (4) was stored under argon atmosphere over silica gel in a desiccator.
- Silicon wafers (t = 0.5 mm) were purchased from Simat, Germany.
- Silicon disks (t = 15 mm, d = 100 mm) for neutron reflectivity experiments were purchased from Crystec, Germany.
- MC sensor arrays were purchased from Octosensis, Micromotive Mikrotechnik, Germany. Used array consisted of eight individual rectangular cantilevers having an area of 500 × 90 µm², ahving thicknesses of 1 and 2 µm and a pitch of 250 µm.
- Deuterated methanol (d4-MeOH) (99.8 Atom%D, Roth) was used as received
- Deuterated terahydrofuran (d8-THF) (99.5 Atom%D, Acros) was used as received
- Poly-vinyl-methyl-ether (PVME) was received from Polyscience Inc. (Niles, IL, USA) in a solution of 50 vol% water. The polymer was dried by freeze drying. The molecular weight was measured with GPC to be M_n = 66000 g/mol with a PDI = 8.
- The used polystyrene (PS) was prepared by anionic polymerization technique with M_n = 26700 g/mol and polydispersity of PDI = 1.05, as measured with GPC.

Materials

- 4-(3′-Chlorodimethylsilyl)propyloxybenzophenone (Cl-BP), was synthesized according to literature[18]. Allowing carbonyl reductions and in order to obtain only partly active benzophenone surfaces, the used Pt-C catalyst was not removed from the product.
- 4-(3'-triethoxysilyl)propoxybenzophenone (EtOH-BP) was prepared, as described in literature[117] using reagents as purchased and solvents of HPLC purity.
- Extra dry toluene (99.8%, Acros) was used as received.
- All other reagents an solvents (HPLC grade) were used without further purification.

III Global Scattering functions: A tool for GISAXS analysis

III.1 Introduction

In chapter II.2.5 the commonly used, though complex DWBA for the analysis of lateral structures was explained. Using the DWBA it became possible to simulate GISAXS with the help of simulation packages[91, 95]. In such kind of DWBA based simulation software packages idealized form and structure factors are included.

Using such kind of simulation approaches will only be meaningful if additional information on the film system is available, including an exact model of the particle shape and lattice. In cases of film systems composed of typical polydispers colloidal particles, which are not arranged on a grid, the use of well defined form and structure factors becomes questionable.

For this work it was of particular interest to analyze GISAXS data from polymeric domains and sol-gel prepared particle systems. Such systems were typically of amorphous structure with polydispersities of 20 – 60%. For such systems, composed of randomly arranged particles or polymeric domains it becomes very difficult to define exact form and structure factors.

However mean sizes and distances are often well defined. Thus, mean radii of gyrations R_g, fractal dimensions and correlation lengths were obtained from GISAXS data.

In the past comparable weakly correlated colloidal powder systems and polymer solutions were analysed with unified fit approaches in transmission geometry[24, 25, 118, 119]. This global model describes scattering from a specimen in terms of multiple structural levels. It is based on a combination of exponential laws, power laws and Bragg based interference functions and describes fractal scattering objects by their fractal dimensions, radius of gyrations and mean centre to centre distances. Applied to GISAXS, a unified approach would make data analysis of the latter film systems easier compared to simulation approaches. However, before the application of such simplified Born Approximation (BA) based models for the analysis of phase separating polymer films, it is of high importance to discuss the applicability and limits towards DWBA described $q_{\parallel}$ scattering in detail in this chapter.

Global Scattering functions: A tool for GISAXS analysis
-
Introduction

Such considerations combine theoretical discussions with simulations and experimental studies. Comparisons of simulations computed for BA based transmission SAXS and DWBA based GISAXS demonstrate the applicability of the unified formalism to GISAXS. For buried particle systems, where refraction effects cannot be neglected, an experimental range for the incident angle of $2\cdot\alpha_c < \alpha_i < 3\cdot\alpha_c$ is proposed. Experimental GISAXS studies were performed on polydisperse Au particle islands prepared by chemical vapour deposition, as a sample system for polydisperse particles located at a free interface. The practical usefulness of the unified fit approach is further demonstrated for buried particles. The study of a model TiO_2/polymethylmethacrylate (PMMA) hybrid material film demonstrates the applicability of the unified fit approach to buried particle systems. The usefulness of unified fits to unravel new physical phenomena is demonstrated in an investigation of percolating networks in hybrid barrier layers used in solar cell applications[120]. Here the analysis of GISAXS measurements using the Unified Fit approach proves the existence of fractal 3D networks proposed from conductive Atomic Force Microscopy experiments and thus clarifies the physics of charge carrier transport in these systems.

III.2 Theory

III.2.1 Approximation of diffuse scattering by BA and intrinsic limits

In this chapter the approximation of diffuse scattering resulting from GISAXS with transmission SAXS theories is discussed.

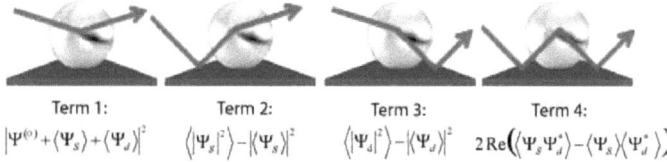

Term 1: $|\Psi^{(e)} + \langle\Psi_s\rangle + \langle\Psi_d\rangle|^2$ Term 2: $\langle|\Psi_s|^2\rangle - |\langle\Psi_s\rangle|^2$ Term 3: $\langle|\Psi_d|^2\rangle - |\langle\Psi_d\rangle|^2$ Term 4: $2\operatorname{Re}(\langle\Psi_s\Psi_d^*\rangle - \langle\Psi_s\rangle\langle\Psi_d^*\rangle)$

Figure III.1: The four possible scattering and reflection effects in the DWBA

As explained in chapter II.2.5.2 diffuse scattering resulting from GISAXS can be best approximated with the DWBA. The DWBA regards combinations of scattering, refraction and reflection effects with the specimen. As shown in chapter II.2.5.2, the DWBA can be explained with four major terms. In contrast BA based transmission SAXS is described by interferences of single scattered waves.

Hence, it has to be shown that under certain conditions the complexity of DWBA scattering can be reduced to single particle scattering. As described in chapter II.2.5.2, *Term 1* describes the scattered intensity at the specular beam position. *Term 2* and *Term 3* describe diffuse scattering from the surface roughness and from interior density fluctuations within the medium, respectively. *Term 4* includes possible correlations between scattered waves from surface roughness and density fluctuations and can in many cases be approximated to 0[89].

Choosing an experimental setup with $2\alpha_c < \alpha_i < 3\alpha_c$, minimization of refraction effects and separation of scattering at the specular beam position and non-specular, diffuse scattering at the Yoneda peak can be achieved. Thus it becomes possible to treat term one independently from term two and three[89, 121]. However, non-specular scattered intensity is still a sum of scattered intensities resulting from surface roughness and interior density fluctuations.

Thus, from $q_\parallel$ GISAXS scattering from surface roughness and interior density fluctuations cannot be separated intrinsically. Using the differential scattering cross section derived from Sinha et al. [75]

$$\frac{d\sigma}{d\Omega} = \frac{Ak_c^2}{(4\pi)^2}|T^iT^f|^2 \Gamma(q) \qquad (\text{III.2.1})$$

and the lateral approximation from Rauscher et.al. [89]

$$\Gamma(q) = \int d^2R_\parallel \exp(-iq_\parallel \cdot R_\parallel) C_w(R_\parallel)t^2 = \hat{C}_w(q_\parallel)t^2 \qquad (\text{III.2.2})$$

one can describe the differential cross section in GISAXS with particle form factors (See chapter II.2.5.2 for details). An example for lateral approximated scattering objects is the upstanding cylinder, which can be described by

$$\hat{C}_w^{Sphere}(q_\parallel, R) = \left[\frac{2\pi R}{|q_\parallel|}J_1(|q_\parallel|R)\right]^2 \qquad (\text{III.2.3})$$

It has to be noted, that the lateral approximation in Eq. (III.2.2) is only valid for cylindrically symmetric scattering objects. From $\hat{C}_w$ of spherical particle islands [91]

$$\hat{C}_w^{Sphere}(q, R) = \left[4\pi R^3 \frac{\sin(qR) - qR\cos(qR)}{(qR)^3}\right]^2 \qquad (\text{III.2.4})$$

one can see that GISAXS in the $q_\parallel$ detector plane has a non constant $q_\perp$ scattering proportion. Consequently the scattered intensities of $q_\parallel$ detector scans at $\alpha_f = \alpha_c$ is not shifted by a constant factor with varying α_i, and β (Eqs. (II.2.22)(II.2.23)(II.2.25)). Nevertheless, simulations presented in chapter III.2 demonstrate that a lateral approximation for $\hat{C}_w^{Sphere}(q, R)$ leads to errors below 20% for $2\cdot\alpha_c < \alpha_i < 3\cdot\alpha_c$, which is sufficient for the analysis of typical colloidal particles.

Apart from such idealized particle systems, density fluctuations in films or on surfaces are in most cases coupled with a certain roughness at the surface or between the interfaces of two phases of different scattering length densities. It was shown (Figure III.1) that diffuse scattering from density fluctuations cannot be decoupled from scattering caused by roughness. Therefore an expression is

Global Scattering functions: A tool for GISAXS analysis - Theory

needed, which includes scattering from rough interfaces coupled with scattering from objects with different scattering length densities.

There are in principle two ways how to describe scattering from rough surfaces. On the one hand scattering from rough surfaces with Gaussian statistics can be described with a height-height correlation function $C_h(|r_{\|}' - r_{\|}''|) = \langle h(r_{\|}')h(r_{\|}'')\rangle$ (see chapter II.2.4)[63]. In this way statistical root mean square (*rms*) values are usually obtained analysing reflectivity data by integrating over the illuminated sample spots. On the other hand, when lateral roughness correlations at $q_{\|} \neq 0$ are to be studied, characteristic power laws can be used to describe the fractality or self affinity of the rough surface or interface (II.2.4). A self-affine rough particle island of an arbitrary shape, which can be supported on a film or buried in a film (e.g. polymer) matrix, is considered. In these cases it may become appropriate, to associate the cut-off length of this rough structure with its radius or radius of gyration R_g.

III.2.2 Unified Exponential/Power-Law Fit model

At this point an expression is needed, which is capable of describing simultaneously scattering from rough interfaces and density fluctuations. In this context density fluctuations are described with different idealized shapes. Beaucage introduced in a set of publications[24, 25, 122] a general unified fit model, which is able to describe scattering over several orders of magnitude for spherical averaged particles in transmission geometry using the BA. This model describes material microstructures in terms of structural levels. Thus it can be applied to model the system's structural features starting from the smallest structural level, such as a nanoparticle towards clusters of particles up to the macro-scale. It was applied successfully in transmission scattering geometry to several particle systems[123,124,125]. In addition, it was also applied to scattering from soft matter systems[122, 126,127]. Nevertheless, this global model was originally developed for analysis of scattering results in transmission geometry using BA. Therefore it has to be shown that it is (within certain limitations) also applicable to the analysis of scattering problems in grazing incidence geometry.

For one structural level the scattered intensity in the unified fit approach is given by

Global Scattering functions: A tool for GISAXS analysis - Theory

$$I(q) = G \exp\left(-\frac{q^2 R_g^2}{3}\right) + B\left[\frac{\left(\mathrm{erf}\left(qR_g/\sqrt{6}\right)\right)^3}{q}\right]^P \qquad (\mathrm{III.2.5})$$

The first term corresponds to Guinier's law and describes the size of spherical averaged particles. It is related to the radius of a sphere by $R_g = \sqrt{\frac{3}{5}} R$. For upstanding particles with cylindrically symmetries, q can be substituted with $q_{\parallel}$ and term one in Eq. (III.2.5) with $G \exp\left(-\frac{q_{\parallel}^2 R_c^2}{2}\right)$. R_C is defined as the radius of gyration of the cross-sectional area of the particle and can be related to the radius of the cross-sectional area by $R_C = \frac{1}{\sqrt{2}} R_{\parallel}$.

The second term in Eq. (III.2.5) corresponds to the structural limited Porod regime, with the Porod prefactor B. The cubed error function limits the fractal regime of the structure's surface at low q over three possible orientations with its radius of gyration R_g. For smooth spherical particles q^{-4} dependence is obtained, while for surface scattering from cylindrically symmetrical structures q^{-3} dependence is obtained[128]. Such power law decays can be found for various fractal geometries as discussed in the following subsection.

For $q_{\parallel}$ scattering from cylindrically symmetric particles Eq. (III.2.5) can be rewritten with a squared error function allowing a power law cut-off over two orientations

$$I(q_{\parallel}) = G \exp\left(-\frac{q_{\parallel}^2 R_C^2}{2}\right) + B\left[\frac{\left(\mathrm{erf}\left(q_{\parallel}R_g/\sqrt{6}\right)\right)^2}{q_{\parallel}}\right]^P \qquad (\mathrm{III.2.6})$$

When more than one structural level is present in a sample, the unified scattering intensity described in Eq. (III.2.5) can be extended to[24]:

Global Scattering functions: A tool for GISAXS analysis - Theory

$$I(q) \approx \sum_{i=1}^{n} G_i \exp\left(\frac{-q^2 R_{gi}^2}{3}\right) + B_i \exp\left(\frac{-q^2 R_{g(i-1)}^2}{3}\right) \left[\frac{\left(erf\left(qR_{gi}/\sqrt{6}\right)\right)^3}{q}\right]^P \quad \text{(III.2.7)}$$

Eq. (III.2.7) is a sum of scattered intensities over n structural levels. When there are correlations between two structural levels, as in particle aggregates, a second term, which limits Porod scattering for $n > 1$ at $R_{g(i-1)}$ has to be introduced. For $q_\parallel$ scattering of cylindrically symmetric structures an analogue expression can be written

$$I(q_\parallel) \approx \sum_{i=1}^{n} G_i \exp\left(\frac{-q_\parallel^2 R_{ci}^2}{2}\right) + B_i \exp\left(\frac{-q^2 R_{c(i-1)}^2}{2}\right) \left[\frac{\left(erf\left(q_\parallel R_{gi}/\sqrt{6}\right)\right)^2}{q_\parallel}\right]^P \quad \text{(III.2.8)}$$

In SAXS analysis the Guinier and Porod prefactors G and B, can be related to the Polydispersity Index (PDI) of spherical averaged particles in SAXS by $PDI = BR_g^4/(1.62G)$ [129]. However, simulations presented later show that such PDI approaches are not valid when applied to $q_\parallel$ scans in GISAXS at $\alpha_i > \alpha_c$, because of the non constant scattered $q_\perp$ portion to $\hat{C}_w^{Sphere}$ (Eq.(III.2.4)). Therefore it is not recommended to use such kind of BA based PDI approaches to estimate particle polydispersity in GISAXS for $\alpha_i > \alpha_c$.

III.2.3 Fractal objects

According to Eq. (III.2.5) power law decays are observed in the high q regimes. The power law exponent P is a characteristic value for the fractality of the probed fractal object. For surface fractal objects the scattered intensity in the high q regime can be approximated as

$$I(q) \propto M^2(qR)^{-\alpha} \propto S \quad \text{(III.2.9)}$$

where, M is the total mass, which scales with R, and S is the surface of a scattering body. For three

Global Scattering functions: A tool for GISAXS analysis
Theory

dimensional objects M scales with R^3, while for two dimensional objects M scales with R^2. S scales with R^{D_S}. For a smooth two dimensional surface $D_S = 2$, while for a one dimensional cross sectional circle $D_S = 1$. Following Eq. (III.2.9) can be rewritten to

$$I(q) \propto R^6 (qR)^{-\alpha} \propto R^{D_s} \qquad (\text{III.2.10})$$

for three dimensional scatterers, such as spheres. Since the exponent of R has to be equal on both sides, one obtains $\alpha = 6 - D_S$ and $I(q) \propto q^{-6+D_S}$. Following the notations from Eq. (III.2.5) and under exclusion of the limiting *erf* one obtains the Porod law for ideal spherical particles $I(q) = Bq^{-4}$. For fractal surfaces $D_S = D - h$, where D is the topological dimension and h the Hurst parameter, as introduced in chapter II.2.4, $I(q) = Bq^{-(4-h)}$ is obtained.

For two dimensional scatterers such as boundaries of upstanding cylinders, α is defined as $\alpha = 4 - D_S$. With $D_S = 1$ one obtains $I(q) = Bq^{-(3-h)}$ for the Porod decay.

For high q scattering of particle clusters power law decays can be explained by scattering from mass fractal objects. The mass M of a mass fractal object scales with R^{D_M}. The mass fractal dimension is defined as $D_M = D - h$.

Accordingly, Eq. (III.2.9) can be rewritten to:

$$I(q) \propto M \left(qM^{\frac{1}{D_M}} \right)^{-\alpha} \propto q^{-\alpha} M^{\left(1 - \frac{1}{D_M}\alpha\right)} \qquad (\text{III.2.11})$$

In order to obtain a result independent of the mass one obtains $I(q) = Bq^{-D_M}$. For high q scattering from polymeric mass fractals the intensity decay can be written as a function of the degree of polymerization N and R_g:

$$I(q) \propto N(qR_g)^{-\alpha} \propto N(qN^\nu)^{-\alpha} = q^{-\alpha} N^{(1-\alpha\nu)} \qquad (\text{III.2.12})$$

with the polymeric scaling exponent ν. For independency of Eq.

(III.2.12) from N, $\alpha = \dfrac{1}{\nu}$ is obtained. In Table III.2.1 the scaling exponents for most common chain models are listed.

Table III.2.1: Scaling exponents for different chain models

	ν	α
Gaussian chain	1/2	2
rod like chain	1	1
three dimensional excluded volume chain	3/5	5/3
two dimensional excluded volume chain	3/4	4/3

III.2.4 Weakly correlated systems

Transverse scattering data from nanostructured film systems often exhibit interference peaks. A description of these interference peaks has been included into the general unified description of scattering from multi length scale systems[24]. The scattered intensity for a correlated system is given by:

$$I_{corr}(q) = I(q)S(q) \qquad (III.2.13)$$

where the term $I(q)$ corresponds to the scattered intensity for a non-correlated system (i.e. the intensity given in Eqs. (III.2.7), (III.2.8)) and $S(q)$ is the factor accounting for correlations between the scattering domains. The interference function $S(q)$, included in Eq. (III.2.13), is used to modify only the two terms of a single structural level, while the remaining levels are not affected. Based on Born-Green theory, $S(q)$ is a function, which describes the correlation between colloidal particles or domains in terms of a radius of correlation ξ and a packing factor κ:[118, 119]

Global Scattering functions: A tool for GISAXS analysis - Theory

$$S(q) = \frac{1}{1 + \kappa F(q,\xi)} \quad \text{with} \quad \kappa = 8 \frac{v_0}{v_1} \qquad (\text{III.2.14})$$

The packing factor κ describes the degree of correlation ($0 \leq \kappa \leq 5.92$) and $F(q,\xi)$ is the 'form factor' for structural correlations occurring at an average distance ξ [118]

$$F(q,\xi) = 3 \frac{\sin(q\xi) - q\xi \cos(q\xi)}{(q\xi)^3} \qquad (\text{III.2.15})$$

The packing factor κ is proportional to the ratio of the average volume of a domain occupied by a material with certain density in respect to the average total available volume of the domain. This is indicated by a volume ratio v_0/v_1, which reaches a maximum of 0.74 for hexagonal or cubic close-packed crystal structures. Multiplying this maximum of v_0/v_1 with the factor 8 (Eq. (III.2.14)), yields $\kappa_{max} = 5.92$. Eq. (III.2.14) describes the scattering amplitude of a sphere. The spherical function is convenient since it can be easily interpreted in a physical sense that it is an average distance of correlation. For the analysis of correlated particle islands arranged in a low correlated 2-D particle grid it becomes appropriate to calculate v_0/v_1 in respect to a 2-D system. Accordingly the volume ratio v_0/v_1 can be exchanged with an area ratio a_0/a_1. For a perfect 2-D crystal $a_0/a_1 = 0.78$ and $\kappa_{max} = 6.24$ is obtained. One has to keep in mind that Eqn. ((III.2.13), (III.2.14), (III.2.15)) are only applicable to describe correlations, when the form factor of a structure can be decoupled from its interference function. This is typically the case for $\kappa < 4$.

Global Scattering functions: A tool for GISAXS analysis - Experimental verification

III.3 Comparison with Simulations

As a first proof for the adaptability of the unified fit approach to $q_\parallel$ dependent GISAXS, $q_\parallel$ scattering graphs for spherical and cylindrical particle shapes were simulated. Comparisons of BA and DWBA allows to compare the unified fit for transmission SAXS with GISAXS. Various IsGISAXS simulation results using BA and DWBA are compared with calculated intensity from the unified fit model using the free accessible Irena software[130,v] for Igor Pro, Wavemetrics Inc.
Therefore as a first step simulations were performed, which mimic diffuse surface scattering from monodisperse, smooth particle islands supported on homogenous surfaces. Such simulations mimic experimental GISAXS from Au particle islands as presented in chapter III.4.1. Isotropic spherical particles were simulated with $R^{sim} = R_x = R_y = R_z = 5$ nm (Figure III.2a,b). Particle islands of cylindrical symmetries were simulated with $R_\parallel^{sim} = R_x = R_y = 5$ (Figure III.2c,d). R_g^{Sim} and R_c^{Sim} were calculated according to $R_g^{Sim}(Sphere) = \sqrt{3/5} R^{Sim}(Sphere)$ and

$R_c^{Sim}(Cylinder) = \frac{1}{\sqrt{2}} R_\parallel^{Sim}(Cylinder)$.

Detailed parameter information included in IsGISAXS simulations is found in Appendix VII.3.1. For BA simulated $q_\parallel$ scans incident and exit angles α_i and α_f were set to $0°$, respectively, which reflects transmission SAXS geometry. For DWBA simulated $q_\parallel$ scans the incident angle was set to $\alpha_i = 0.7°$, equal to the performed GISAXS experiments on the studied Au film, while the exit angle was set to $\alpha_f = \alpha_c (Au) = 0.50°$.
Simulated $q_\parallel$ scans of spherical particle islands were compared with three dimensional averaged intensities from Eq. (III.2.5), while simulated $q_\parallel$ scans of cylindrical were compared with radial averaged intensities from Eq. (III.2.6). The observed power-law decays of q^{-4} for spherical particle islands and of q^{-3} for radial averaged cylindrical particle islands can be addressed to Porod surface scattering as described in the theoretical discussion.
Results of R_g^{Fit} and R_c^{Fit} from Guinier fits (Table III.3.1) follow the employed parameters R_g^{Sim} and R_c^{Sim} with deviations < 10% for BA simulated transmission SAXS and DWBA simulated GISAXS. Additional comparison of Figure III.2a with Figure III.2b shows that oscillations of simulated $q_\parallel$ scans vary from BA to DWBA. This can be related to the q_z dependence of the spherical form factor

---
[v] www.usaxs.xor.aps.anl.gov/staff/ilavsky/irena.html

Global Scattering functions: A tool for GISAXS analysis
Experimental verification

(Eq. (III.2.4)) and a coherent interference of reflected, refracted and scattered waves. However, the obtained R_g^{Fit} value still matches in good agreement with R_g^{Sim}. In contrast to R_g and R_c the obtained ratios of B/G appear to be highly dependent on the scattering geometry and the applied perturbation theory (e.g. BA and DWBA). Consequently in GISAXS analysis at $\alpha_i > \alpha_c$ BA based PDI approaches, which are commonly used in SAXS analysis[129] should be avoided.

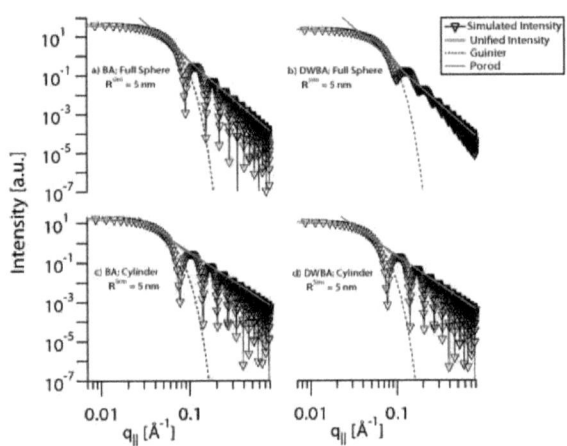

Figure III.2: Comparison of $q_\parallel$ scans from IsGISAXS simulated intensity using BA and DWBA with calculated unified fit intensity using different particle island geometries a,b) Monodisperse full sphere ($R^{sim} = 5$ nm; $R_g^{Sim} = 3.87$ nm); c,d) Monodisperse cylinder ($R_\parallel^{sim} = 5$ nm; $R_c^{Sim} = 3.53$ nm)

Table III.3.1: Unified fit results for particle types simulated in Figure III.2 using BA and DWBA

	Full Sphere		Cylinder	
	BA	DWBA	BA	DWBA
G	42.7	23.4	17.0	11.3
R_g [Å]	41.5	38.0		
R_c [Å]			37.7	37.8
B	5.6·10⁻⁵	3.0·10⁻⁵	3.5·10⁻⁴	2.2·10⁻⁴
P	4	4	3	3

Global Scattering functions: A tool for GISAXS analysis - Experimental verification

Due to refraction effects one should test dependencies on incident angles for buried particle systems in more detail, than for spherical particle islands at the free interface. Based on experimental studies diffuse scattering of ideal spherical TiO_2 particles with $R = 5$ nm buried in a film matrix of a lower density were simulated. Simulated $q_\parallel$ detector scans were compared with the scattered intensity from Eq. (III.2.6) (Figure III.3) for a wide range of α_i. Such testing allows estimating angular ranges in which experimental setups were chosen. Detailed parameter sets can be found in Appendix VII.3.2.

From Figure III.3 one can see that fitted R_g values are in a range of ± 20 % from the theoretical R_g for $\alpha_c \leq \alpha_i \leq 3\,\alpha_c$. It was observed that fits can be misleading for $\alpha_i > 3.0\,\alpha_c$. For the lower limit of the incident angle $\alpha_i > 2.0\,\alpha_c$ is suggested. At smaller incident angles refraction effects lead to an enhanced error in resulting R_g values. As a combined result incident angles in the range of $2\,\alpha_c \leq \alpha_i \leq 3\,\alpha_c$ for buried particle systems are suggested. In this region particle dimensions can be analysed with deviations < 20%, which is sufficient for most polydisperse colloidal particle and polymeric systems.

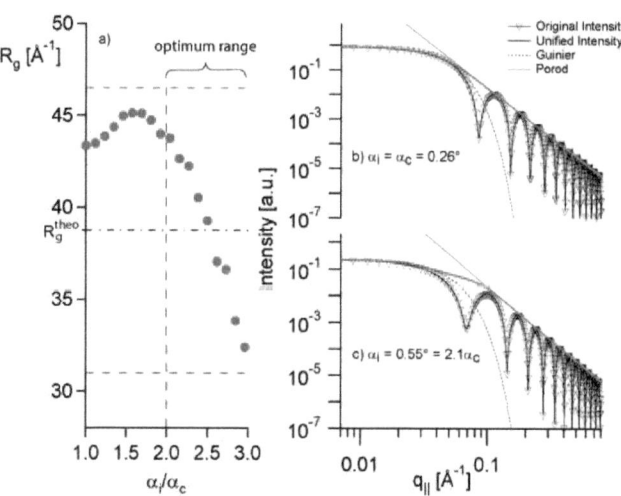

Figure III.3: Unified fit results on monodisperse spherical TiO_2 particles ($R^{Sim} = 5$ nm; $R_g^{Sim} = 3.87$ nm) buried in a film matrix; a) Fitted R_g vs. α_i/α_c; horizontal dashed lines represent range of 20% deviation from the theoretical R_g^{theo}; perpendicular dashed line sets the proposed experimental range of $2\,\alpha_c \leq \alpha_i \leq 3\,\alpha_c$ b) example of fit at $\alpha_i = 0.26° = \alpha_c(TiO_2)$; c) example of a fit at $\alpha_i = 0.55° = 2.1\,\alpha_c$.

III.4 Experimental verification

III.4.1 Unified analysis from model systems

Figure III.4 shows experimental $q_\parallel$ scattering data, obtained from GISAXS pattern at $\alpha_f = \alpha_c$. Studying experimental data transverse scattering from two structural levels, which are separated by a correlation peak, can be observed. The found Porod decay of ~ 4 at high q suggests the presence of polydisperse hemispherical particle islands. Therefore, the use of the three dimensional averaged Eq. (III.2.7) can be used. From fit results of the 1st structural level, an average lateral dimension of the Au particles of $R_g^{\,1} = 11$ nm can be assigned.

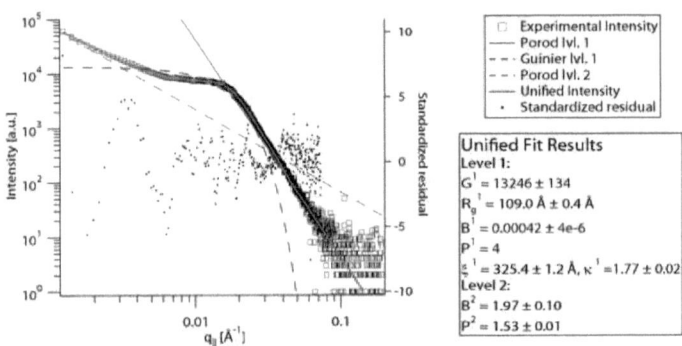

Figure III.4: Experimental GISAXS data of low correlated Au film prepared by CVP fitted with Eqn. ((III.2.7), (III.2.13), (III.2.14), (III.2.15)).

Results from correlation functions (Eqn. (III.2.13), (III.2.14), (III.2.15)) included in the applied fit suggest that the covering particle islands are randomly arranged - this can be deduced from the low packing factor $\kappa^1 = 1.8$ - with a mean centre to centre distance of $\xi^1 = 33$ nm. In the experimental $q_\parallel$ range it was able to record only few data-points, which account for Porod scattering of level 2. However, the found power law decay of $P^2 \approx 1.5$ can give evidence to a mass fractal, which is related to a two dimensional arrangement of Au particle islands.

Global Scattering functions: A tool for GISAXS analysis
-
Experimental verification

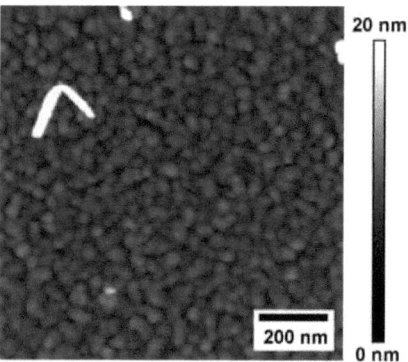

Figure III.5: AFM image of the studied 20 nm thick Au film on a silicon wafer.

GISAXS $q_\parallel$ scans from the studied Au film and corresponding unified fit results were compared with SPM data from randomly distributed Au islands (Figure III.5). Drawing on the assumption of hemispherical islands, mean particle island sizes of D = 26 nm with a polydispersity of 27 % can be assigned. This is in good agreement with the obtained R_g^1 fit value from GISAXS analysis. For islands centre to centre correlation lengths mean values of ξ = 32 nm ± 10 nm were obtained. The average particle island distance of 32 nm with the calculated standard deviation of 32% is in good agreement with the average distance and the low degree of correlation κ, found by GISAXS experiments. To summarize these GISAXS and SPM results it can be concluded that predictions within a reasonable accuracy on particle island sizes, forms and arrangements including mass fractal dimensions can be made with the unified fit approach, when transverse $q_\parallel$ detector scans at $\alpha_f = \alpha_c$ are performed.

In order to experimentally test the formalism for GISAXS on buried particle systems $q_\parallel$ scans deduced from GISAXS in combination with comparative SEM results from spin-coated TiO_2 (Solaronix T, BASF)/PMMA hybrid material films are studied. Figure III.6 shows a $q_\parallel$ detector scan at $\alpha_f = \alpha_c$ extracted from an obtained GISAXS pattern at α_i = 0.55°. At high $q_\parallel$ a power law decay of $P^1 \approx 4$ for diffuse Porod scattering of the first structural level, meaning the immersed TiO_2 particles, with R_g^1 = 9 nm can be found.

Global Scattering functions: A tool for GISAXS analysis - Experimental verification

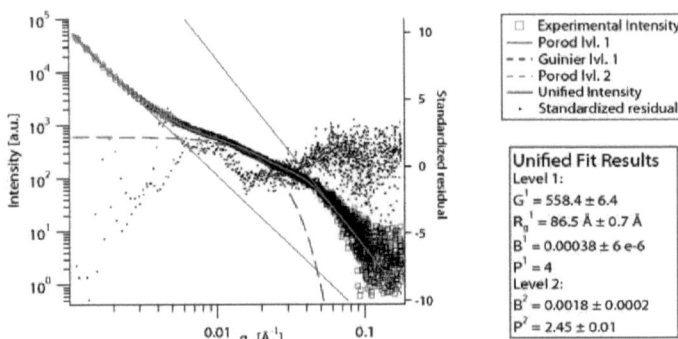

Figure III.6: Experimental GISAXS data of uncorrelated spin coated TiO$_2$/PMMA film fitted with Eq. (III.2.7).

At high $q_\|$ a power law decay of $P^1 \approx 4$ for diffuse Porod scattering of the first structural level, meaning the immersed TiO$_2$ particles, with $R_g^1 = 9$ nm can be found. Immersed TiO$_2$ particles seem to be uncorrelated and far separated in the PMMA matrix. This led to negligibly small contributions of the interference function $S(q_\|)$ to the unified intensity and was therefore approximated as unity. From the 2nd level Porod approach a power law decay in the range of $2 < P^2 < 3$ was found. This finding proposes none surprisingly a particle arrangement in an arbitrary three dimensional mass fractal. From comparison with SEM analysis (Figure III.7) a mean particle size of $D = 18$ nm with 20% polydispersity was assigned. This result is again in reasonable agreement with the found radius of gyration from GISAXS analysis of the TiO$_2$/PMMA film.

Global Scattering functions: A tool for GISAXS analysis - Experimental verification

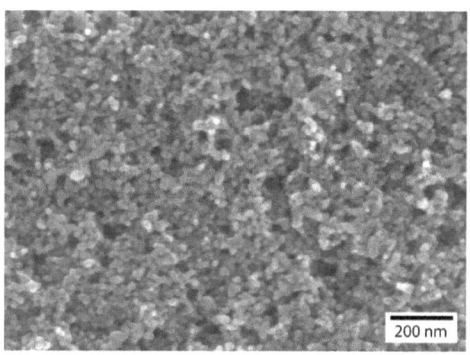

Figure III.7: Scanning electron micrograph of the TiO$_2$ particles (Solaronix T)

III.4.2 Unified analysis from novel TiO$_2$/(PEO)MA–PDMS–MA(PEO) films

In this last experimental verification the usefulness of the unified fit approach for clarifying morphologies in functional films containing polydisperse colloidal particles is demonstrated. Using conductive SPM it was suggested that electrical charges are transported through partly ceramized percolating TiO$_2$/(PEO)MA–PDMS–MA(PEO) networks with applications in solar cells[120]. A ceramized isolating PDMS shell prevents lateral shortcuts. Only from SEM studies (Figure III.9) the morphology of the film becomes not clear. Thus no clear conclusion on the charge transport mechanism could be made without comparative results. In Figure III.8 the analysed experimental GISAXS intensity of a 20 nm thick TiO$_2$/(PEO)MA–PDMS–MA(PEO) film, is shown. The average R_g of the primary uncorrelated TiO$_2$ particles is found to be 4.0 nm. From $P^2 = 2.2$ it can be concluded that the small particles are not two dimensionally arranged in the film matrix, but in a three dimensional mass fractal. The charge transport mechanism can be specified by electrons percolating perpendicular to the film surface through a network of TiO$_2$ particles. Particles are not correlated laterally. Therefore lateral shortcuts are prevented.

Global Scattering functions: A tool for GISAXS analysis - Experimental verification

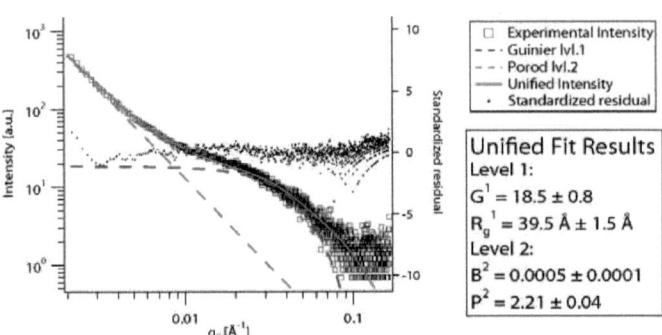

Figure III.8: Experimental GISAXS data of spin coated $TiO_2/(PEO)MA–PDMS–MA(PEO)$ film fitted with Eq. (III.2.7).

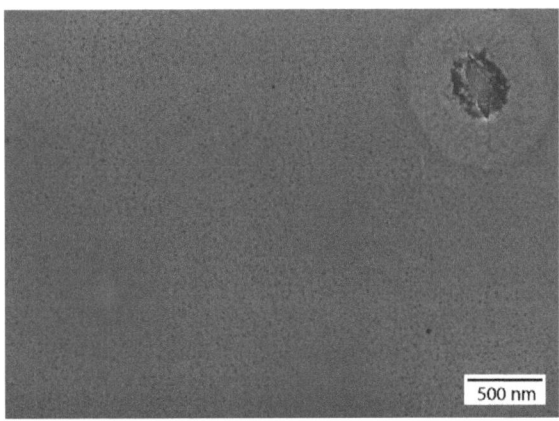

Figure III.9: Scanning electron micrograph of the $TiO_2/(PEO)MA–PDMS–MA(PEO)$ film

III.5 Summary

In this part of the thesis was shown that the application of versatile global scattering functions – such as the unified fit equations - which were used over ten years for data analysis in transmission SAXS, USAXS and SLS, to transverse $q_\parallel$ detector scans from GISAXS experiments is allowed within certain limits. Theoretical consideration and simulations show that the unified formalism can be used for particles with spherical or cylindrical symmetry, either placed at the free surface or buried in a film matrix within deviations < 20 % from theoretical values. Such theoretical deviations are usually not critical for the analysis of typical colloidal particles with polydipersities in the range of 20-60 %. It was concluded from simulations that experimental incident angles should be chosen carefully before conducting the GISAXS experiment. For the studied TiO_2 containing film system best results were obtained for incident angles in the range of $2 \cdot \alpha_c \leq \alpha_i \leq 3 \cdot \alpha_c$. Comparing experimental GISAXS results with particle analysis from SPM and SEM studies it was possible to verify mean particle island sizes and centre to centre correlation lengths. In addition, further considerations on particle arrangements, described by packing factors and mass-fractal dimensions, were shown to be useful. A clear limit to the presented approach is that polydispersity approaches used in transmission SAXS[129], can be misleading for analysis of GISAXS at $\alpha_i > \alpha_c$. Thus, it is not proposed to use such kinds of approaches within a BA based interpretation framework. However, using the unified fit approach for GISAXS on novel functional films can help to unravel physical mechanisms where microscopic techniques like SEM or SPM would not be sufficient. Due to the applicability of the discussed BA based unified fit approach for GISAXS of fractals composed of colloidal particle systems, the unified fit formalism is expanded for arbitrary polymeric mass fractals in chapter IV.4.3.

IV Thermal response of surface grafted two-dimensional PS/PVME blend films

IV.1 PS/PVME bulk properties

Before discussing experimental results from the grafted two-dimensional PS/PVME blend films, the blend's bulk properties have to be addressed. For bulk investigations PS/PVME homopolymer and blend films were prepared from 1% wt. toluene solutions. Films were dropcast on precleaned glass substrates, dried under vacuum and removed from the glass substrate for differential scanning calorimetry (DSC) measurements.

In Figure IV.1 the DSC thermograms of the PS and PVME homopolymer are shown. In both DSC thermograms a negative step in the base line can be observed, which can be attributed to an endothermal second order phase transition from a glassy polymer phase towards a rubbery phase. The glass transition temperature T_g can be assigned by the onset of the heat decay and T_g (PVME) = -29 °C and T_g (PS) = 104 °C are obtained.

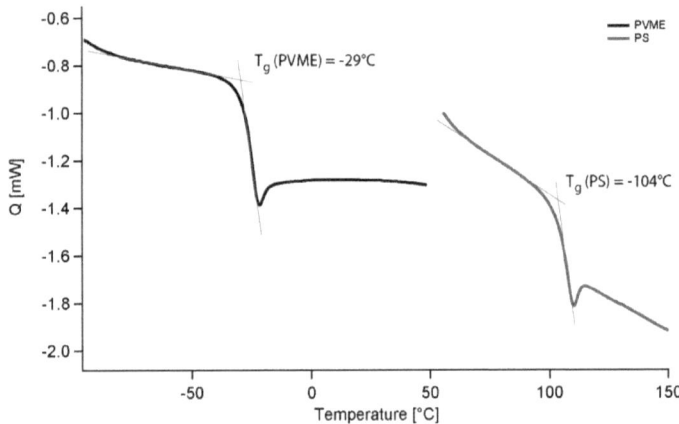

Figure IV.1: DSC thermograms of PS and PVME bulk homopolymer.

Similar step transitions are observed in the thermograms from the PS/PVME = 20/80 and PS/PVME = 40/60 blend (Figure IV.2). In agreement with experimental results from Bank et al.[19] PS/PVME blend films composed of PVME weight fractions ≤ 0.4 have single glass transition

Thermal response of surface grafted two-dimensional
PS/PVME blend films
-
PS/PVME bulk properties

temperatures near the PVME bulk's T_g. However, the width of the step transition increases for increasing PVME weight fractions. In contrast no glass transitions can be observed for PS/PVME = 60/40 and PS/PVME = 80/20 bulk blends. The polymer blends were therefore found in a glassy state for the entire temperature range.

In addition to step transitions resulting from glass transitions, exothermal peaks can be observed in the DSC thermograms at high temperatures for all four polymer blend mixtures. Exothermal peaks can be attributed to first order phase transitions. In this case the exothermal heat can be attributed to a polymer/polymer phase separation from a single mixed phase towards a two phase system (chapter II.3.3). The phase separation temperatures, T_s, are indicated by the positions of the peak maxima. The resulting exothermal heat ΔQ_s can be allocated to the peak integrals.

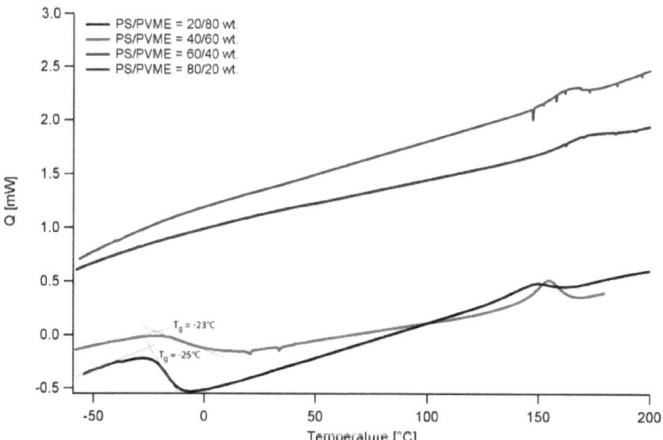

Figure IV.2: DSC thermograms of PS/PVME blend bulk films.

Table IV.1.1: T_g, T_s and ΔQ_s values deduced from DSC thermograms shown in Figure IV.2.

PS/PVME	T_g [°C]	T_s [°C]	ΔQ_s [mW°C]
20/80	- 25	148	2.5
40/60	- 23	154	3.1
60/40	-	147	1.4
80/40	-	169	1.6

Thermal response of surface grafted two-dimensional PS/PVME blend films

PS/PVME bulk properties

From obtained values (Table IV.1.1) it can be seen that highest resulting heats were measured for the PS/PVME = 40/60 blend. However, for experimental studies on the two dimensional surface grafted films the PS/PVME = 20/80 was used, since DSC measurement assured a rubbery state at room temperature. This behavior simplified the coating procedure with the Nano-Plotter^{TM} NP2.0, since film formation of higher uniformity was observed for solution casting of rubbery polymer systems.

IV.2 Thin film phase separation

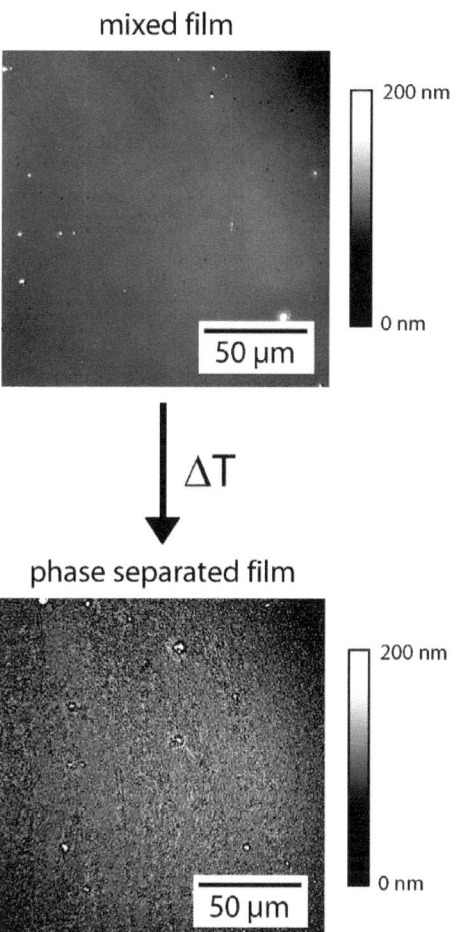

Figure IV.3: Confocal microscope images of a mixed PS/PVME = 20/80 film, which phase separated after annealing at 150°C.

After characterization of bulk film properties, the phase separation behavior of thin spin cast films was studied. The PS/PVME = 20/80 was spin cast at 1000 rpm from a 1%wt. toluene solution. As observed from confocal microscope images uniform mixed PS/PVME were obtained (Figure IV.3). After heating in vacuum for ~ 17h at 150°C, the film's topography was again measured with confocal microscopy. In contrast to the mixed film small domains in the nanometer scale were observed.

IV.3 Preparation of grafted to polymer films

IV.3.1 Introduction to specific and unspecific grafting to routes

Different to the grafting from technique pre-synthesized polymer chains are irreversibly bound to a functionalized surface within the grafting to technique[131]. Traditionally pre-synthesized polymers carry functional groups at specific chain positions, which react irreversibly with a surface immobilized linker molecule. In such way only small amounts of polymer can be immobilized to the surface. Once the first polymer molecules are bound to some active surface sites, they can spread laterally, because of low kinetic hindrances towards their next neighbors. In the following additional polymer chains have to diffuse through the first adsorbed and bound layer. Such diffusion is kinetically unfavored. Thus, active surface sites are hard to access for further polymer immobilization. Grafted to prepared films have intrinsically in common that they are usually only a few nanometers thick[132].

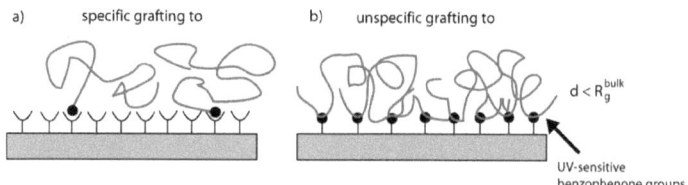

Figure IV.4: Schematic comparison of a) specific grafting to with b) unspecific grafting to using UV-sensitive linkers. In both prepared films film thicknesses are found to be smaller than the bulk's radius of gyration.

Prucker et al.[18] developed a benzophenone derivate linking molecule, which can be activated to a radicalic form with UV. Such radicalic activated linker molecules are able to chemically immobilize unfunctionalized polymer chains by radical substitution mechanisms to the substrate. Depending on the density of benzophenone groups more than one possible grafting point per polymer chain can be introduced. Film thicknesses were found to be similar than for traditional casting to films. However it is suggested within this work that such low film thicknesses are not resulting from non accessible surface sites, but from multiple chain anchoring. All grafting to prepared film samples were prepared according to the unspecific preparation route.

IV.3.2 Surface functionalization with UV-sensitive benzophenone linkers

Three kind of benzophenone (BP) linking film samples were prepared during this work, which varied in the grafting point density. The reaction scheme for the different linker layers is illustrated in Figure IV.5. BP synthesis started from etherification of 4-hydroxybenzophenone with allylbromide[18].

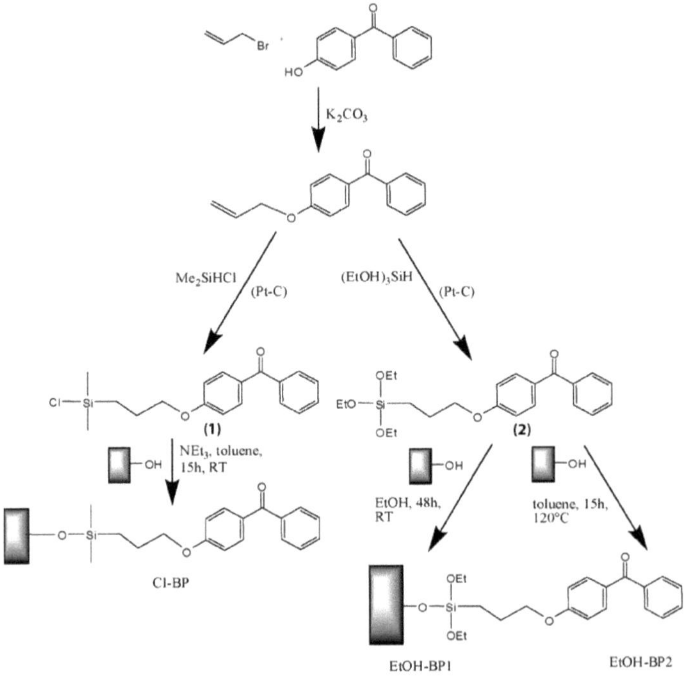

Figure IV.5: Reaction diagram for the preparation of different benzophenone linking layers Cl-BP, EtOH-BP1 and EtOH-BP2.

Thermal response of surface grafted two-dimensional
PS/PVME blend films
-
Preparation of grafted to polymer films

For the preparation of Cl-BP functionalized MC sensor arrays and Si wafer samples 4-Allyloxybezophenone was hydrosilanated with Me_2SiHCl in order to obtain **1**. The hydrosilanization was performed according to literature except for the removal of the Pt-C catalyst. The storage of compound **2** with the catalyst led to fractional reductions of the benzophenone group leading to small grafting point densities (chapter IV.4). Such deactivation of the carbonyl unit by the presence of the Pt-C catalyst is known in literature[18, 133].

The base cleaned substrates (backside gold protected MC sensor arrays and Si wafer pieces) were immersed in a solution of 0.4 mL freshly distilled NEt_3, 0.2 mL of compound **1** in 25 mL of dry toluene. The coating mixture with a concentration of 0.024 mol/L was stirred under Ar atmosphere for 15 h at room temperature. The coated substrates were removed from the reaction mixture cleaned for 3 h under soxhlet extraction with CH_2Cl_2. Such prepared Cl-BP films were measured to be 4.1 ± 0.1 nm thick with a calculated film density of 0.78 ± 0.01 g/cm³. Such high film thicknesses suggest the formation of multilayers, potentially caused by condensations of the reduced species of **1**. However, polymer blend films grafted to Cl-BP specimen allows polymer-polymer phase separations as discussed in chapter IV.4.

For the preparation of EtOH-BP1 and EtOH-BP2, 4-Allyloxybezophenone was hydrosilanated with $(EtOH)_3SiH$ using a Pt-C catalyst according to literature[117]. Unlike for the preparation of Cl-BP, the catalyst was removed by filtration and compound **2** was obtained. EtOH-BP1 and EtOH-BP2 specimen varied in their grafting point densities. EtOH-BP1 was prepared by immersing the substrates in a 1 mmol/L solution of **2** in EtOH. The coating mixture was kept under Ar atmosphere and was stirred at room temperature for 48h. EtOH-BP2 was prepared by immersing the substrates in a 0.12 mmol/L solution of **2** in water free toluene. The coating mixture was boiled for 12 h at 120°C under Ar atmosphere. EtOH-BP1 and EtOH-BP2 samples were thoroughly rinsed with EtOH at the end of the coating process.

EtOH-BP1 and EtOH-BP2 films were measured to be ~ 1 nm thick. Therefore monolayer formation is assumed.

Thermal response of surface grafted two-dimensional
PS/PVME blend films
-
Preparation of grafted to polymer films

IV.3.3 Functionalization with polymers

All grafted polymer films were prepared from polymer solutions dissolved in toluene (1 wt.%). Different homo and blend polymer films were grafted to BP pre-functionalized wafer samples via spincoating at 3000 rpm. The spin coated samples were irradiated at $\lambda = 365$ nm with a total energy of 6.28 J/cm^2 (Süss Micro Tec Delta 80). Excess polymer was removed with soxhlet extraction using CH_2Cl_2 for 12 h.

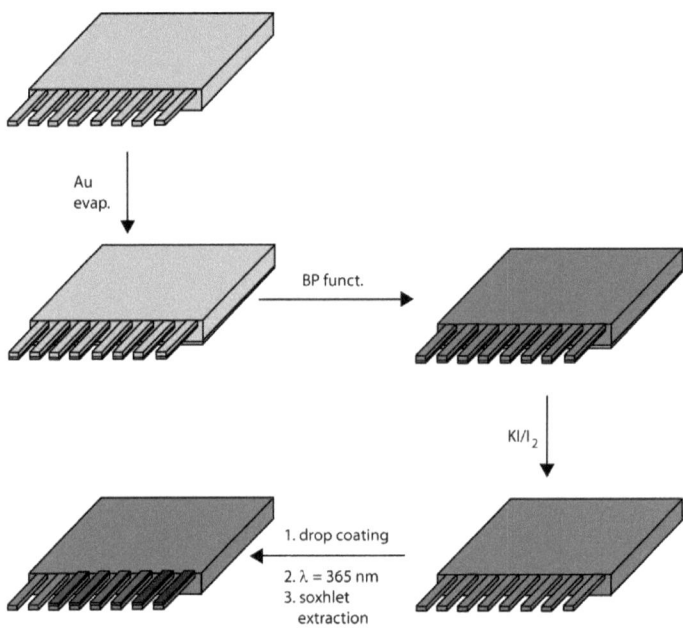

Figure IV.6: Preparation Scheme of MC sensor arrays: The arrays backsides were passivated were passivated with a Au layer; In the following step the arrays were functionalized with the UV sensitive BP linker. After functionalization the protecting Au layer was removed with KI/I2 solution. Grafting to of individual polymers was performed by selective drop casting followed by UV irradiation and soxhlet extraction.

MC sensor array specimen had to be coated selectively on the MC sensors topside, which was achieved by backside protection with Au (chapter II.5.2). The Au layer was removed with KI/I_2 solution before the polymer grafting step as illustrated in Figure IV.6.

To allow differential bending experiments, individual signals from polymer grafted MC sensors and

Thermal response of surface grafted two-dimensional PS/PVME blend films

Preparation of grafted to polymer films

reference MC sensors – such as the MC functionalized only with the BP – have to be recorded simultaneously. Polymer blend and homo-polymer solutions were selectively applied to single MC sensors using a video-assisted spot deposition technique[45, 46]. Solution droplets were drop cast with a piezo controlled nanoliter pipette, ejecting volumes of ~ 0.4 nL, using a Nano-Plotter© 2.0 device (GeSim, Germany). Using a graphical user interface cross hairs were addressed to the desired droplet positions on the MC sensor array (Figure IV.7a). After spot positioning the polymer solution was taken up with the piezo driven pipette. To each crosshair position one single droplet was applied. Within three coating runs visible polymer films as shown in Figure IV.7b were obtained.

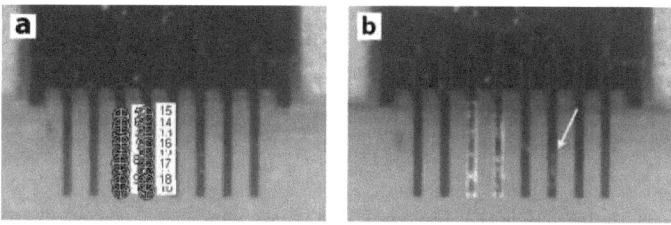

Figure IV.7: Drop casting process with the Nanoplotter2 device. a) Camera image of positioned spots on MC sensors via a graphical user interface; b) Camera image of two MC sensors after three coating runs; unspotted MC sensors are left uncoated.

Obtained crack formation was not estimated to be crucial, because only the first adsorbed polymer layers are bound to the substrate, while excess polymer is rinsed off. In addition bending effects from small visible artifacts observed on some MC sensors (illustrated by the yellow arrow) are assumed to be less important, since they cover only a small area of the whole sensor. In a typical coating process two sensors were coated with the same polymer solution. After drop casting the coated arrays were irradiated at $\lambda = 365$ nm with a total energy of 6.28 J/cm^2 and the first adsorbed polymer layers were chemically immobilized. Following, unbound polymer was removed from the arrays by means of soxhlet extraction with CH_2Cl_2 for ~ 15h.

Compared to MC sensor arrays, the grafting to process for reference Si wafer sample could be simplified. There was no need to introduce a protecting Au layer and instead of drop casting, spin coating using a Süss Micro Tec Delta 80 spincoater could be used. The spinning velocity was usually 3000 rpm with an acceleration of 1000rpm/s.

IV.4 Effect of grafting point densities

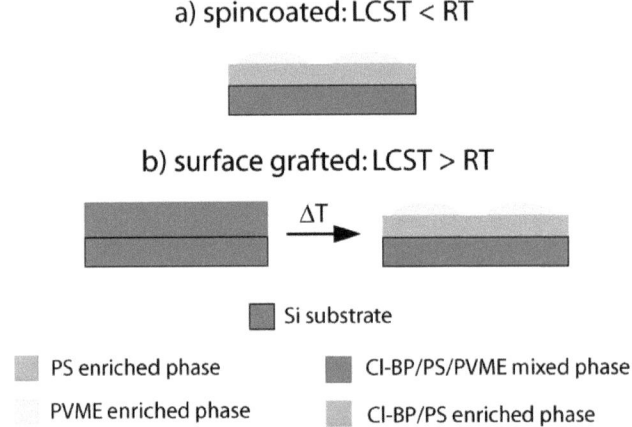

Figure IV.8 a) phase separated two-dimensional PS/PVME film, prepared by spin coating [115]; b) expected dewetting mechanism at T > RT of a constrained, surface anchored PS/PVME film.

This chapter shows that phase separations in two dimensional polymer films can be tuned by employing entropically constraining grafting points. Experimental results on surface grafted two dimensional polystyrene PS/PVME blend films using Cl-BP and EtOH-BP1 pre-functionalized Si surfaces. It is shown that the grafted polymer blends LCST is highly dependent on grafting point densities. Compared to non grafted two dimensional films[115] it was possible to raise the blends LCST above room temperature using low grafting point densities, which were prepared from Cl-BP functionalized surfaces (chapter IV.3.2) . Highly constrained films prepared from EtOH-BP1 did not show polymer-polymer phase separation in the studied temperature range. In addition to in situ structural analysis performed with SPM, µ-GISAXS and µ-XR, surface stress investigations with the use of MC sensor arrays gave detailed insight into the phase separation mechanism. It was concluded that phase separations result in dominating attractive entropic spring mechanisms with opposing repulsive effects resulting from interfacial tensions.

Thermal response of surface grafted two-dimensional PS/PVME blend films
- Effect of grafting densities

IV.4.1 Hypothesis

It is well known that properties of polymer brush systems are highly dependent on their grafting densities[26, 27]. In contrast to e.g. end grafted polymer brush films, polymer chains which are grafted to benzophenone functionalized surfaces are grafted unspecifically at more than one chain segment to the surface (Figure IV.9). Varying grafting densities of the benzophenone functionalized silanes alters the number of polymer/surface grafting points per chain and introduces entropic constraints to demixing polymer blend films. For grafted polymer blend systems with lower LCST, such as the PS/PVME blend, entropic constraints should lead to an increase of the LCST, because the PVME dewets under recovering of conformational entropy in two dimensional films[115](Chapter II.3.3). Hence, increasing numbers of entropic constraints should raise the LCST to much higher temperatures.

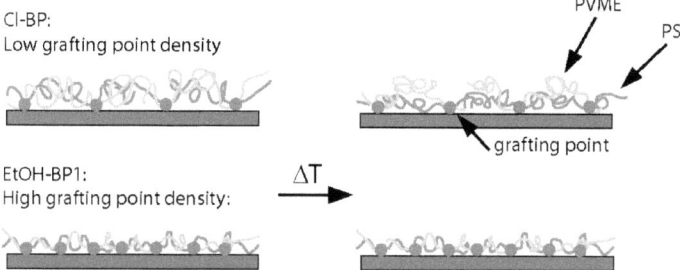

Figure IV.9: Schematic representation of the thermo response of surface anchored two dimensional PS/PVME blend in the case of low (Cl-BP) and high grafting point densities(EtOH-BP); Based on results from Tanaka et al.[115] low grafted blend films showed dewetting of the PVME from the mixed silane/PS phase; in highly grafted/entropically constrained films no dewetting was observed.

IV.4.2 SPM results

As a first proof of the latter discussion *in situ* SPM studies were performed to demonstrate that the low grafted Cl-BP/PS/PVME films have a LCST of RT < T_{LCST} < 150°C, while highly grafted EtOH-BP1/PS/PVME did not show a phase separation effect (Figure IV.10). SPM height images of the Cl-BP/PS/PVME recorded at RT and at 150°C (Figure IV.10a) under inert gas atmosphere from the same specimen showed an increase in roughness from 0.3 nm to 0.6 nm. In contrast EtOH-

Thermal response of surface grafted two-dimensional PS/PVME blend films - Effect of grafting densities

BP1/PS/PVME films showed no changes in roughness changes at 150°C (Figure IV.10b).

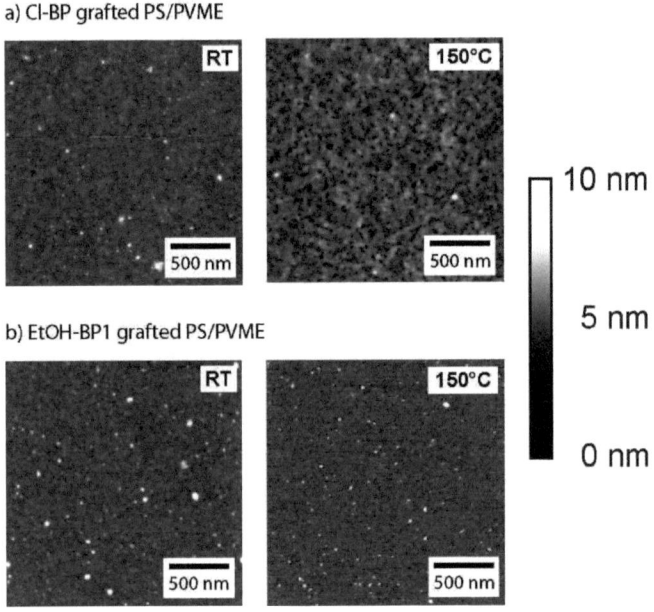

Figure IV.10: SPM height images at RT and 150°C of PS/PVME films grafted to a) Cl-BP prefunctionalized surface; b) EtOH-BP1 prefunctionalized surface

The roughness increase during annealing was attributed to the formation of polydisperse domains with average diameters of ~ 60 nm. The domains showed no long range order, but average centre to centre distances of ~ 110 nm and average domain heights of ~ 2 nm. According to dewetting mechanisms in non anchored films[115] dewetting of the hydrophilic PVME from the miscible hydrophobic Cl-BP/PS phase is assumed. From coarse domain height estimations of ~ 2 nm and on the basis on the work of Shuto et al.[134] one can see, that the vertical chains R_g would be by a factor of ~ 9 lower than the R_g of an unperturbated PVME chain. This implies that the chains are still in a constrained conformation state after phase separation.

Thermal response of surface grafted two-dimensional
PS/PVME blend films
-
Effect of grafting densities

IV.4.3 μ-XR and μ-GISAXS results

For clarification of phase separation mechanisms μ-XR and μ-GISAXS experiments on Cl-BP grafted PS/PVME blend and homo polymer films were performed to obtain more accurate and comparable information from larger sample areas. To be sure that the Cl-BP coating is equal for all the studied films and to ensure equal environmental conditions a coated MC sensor array was used for scattering and reflectivity experiments. It has to be further noted that the μ-XR and μ-GISAXS experiments were performed at the same specimen at one single heating cycle. Such procedure allowed correlating perpendicular film information from μ-XR with lateral film information from μ-GISAXS at RT and 150°C.

Experimental μ-XR data (Figure IV.11) was fitted using Parratt's formalism[135, 136]. Obtained parameters show that Cl-BP and PS films were of equal thickness of 4 nm (Table **IV.4.1**). The estimated stretched length of the Cl-BP molecule is < 1 nm. Thus a kind of silane multilayer structure is observed within the used hydrolysis approach. Equal film thicknesses support the assumption of a hydrophobic mixed Cl-BP/PS phase. Due to the low film densities, the availability of functional benzophenone groups is small enough to allow formations of second, lower densed phases in the grafted PVME and PS/PVME systems. Following, PVME chains seem not to be completely miscible in the Cl-BP or in the miscible PS/Cl-BP phase.

During samples annealing the Fresnel oscillations shifted to higher q_z values. Due to the absence of second Fresnel minima in the experimental q_z range, it was not possible to distinguish between a one or two layer system. However, decreases in film thicknesses combined with decreasing material densities could be obtained from the fits for all studied films. It seems that the Cl-BP multilayer structure is partly collapsing with an accompanied decomposition during the heating process. Nevertheless, results from μ-GISAXS show that such Cl-BP film decompositions are not disturbing the grafted polymers microstructure.

In contrast the EtOH-BP1 anchored PS/PVME films were measured to be 1.5 nm thick, (Figure IV.12), which did not change during heat treatment. Compared to Cl-BP films no bilayer-system with a lower densed polymer phase on top was found.

Thermal response of surface grafted two-dimensional
PS/PVME blend films
-
Effect of grafting densities

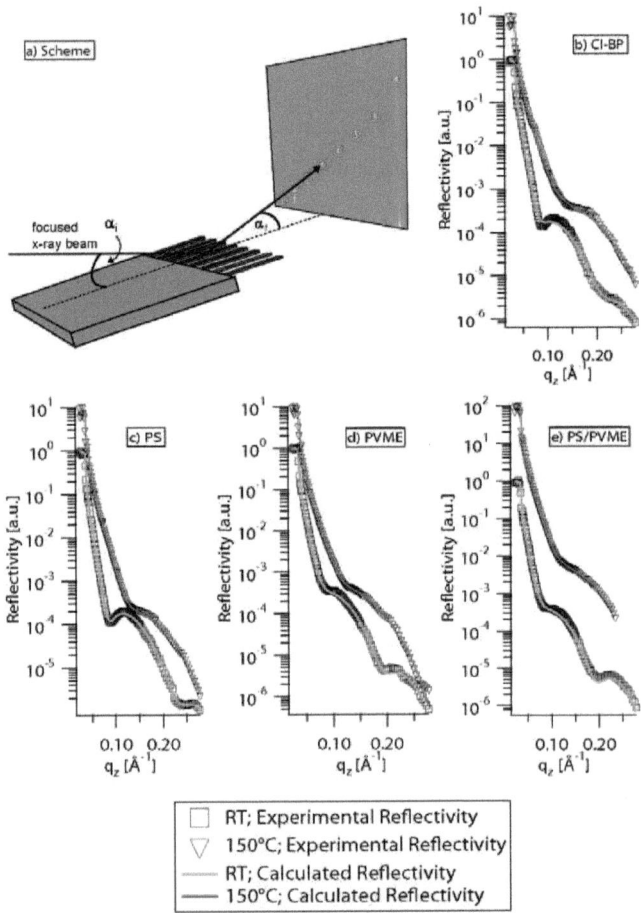

Figure IV.11: a) Scheme of experimental setup; Reflectivity curves of b) Cl-BP MC sensor; c) Cl-BP/PS grafted MC sensor; d) Cl-BP/PVME grafted MC sensor; e) Cl-BP/PS/PVME grafted MC sensor, at RT and 150°C, respectively; reflectivity curves for heat treated specimen are shifted by a factor of 10 for improved visualization.

Thermal response of surface grafted two-dimensional
PS/PVME blend films
-
Effect of grafting densities

Table IV.4.1: Fit results of film thicknesses t and film densities ρ from µ-XR curves (Figure IV.11)

	Cl-silane		PS	
	RT	150°C	RT	150°C
t^1 [nm]	4.1 ± 0.1	2.9 ± 0.2	4.0 ± 0.1	2.9 ± 0.2
ρ^1 [g cm^{-3}]	0.78 ± 0.01	0.45 ± 0.01	0.91 ± 0.01	0.47 ± 0.01

	PVME		PS/PVME	
	RT	150°C	RT	150°C
t^1 [nm]	3.4 ± 0.1	3.2 ± 0.2	3.4 ± 0.1	3.1 ± 0.2
ρ^1 [g cm^{-3}]	0.78 ± 0.02	0.50 ± 0.02	0.90 ± 0.02	0.46 ± 0.01
t^2 [nm]	1.7 ± 0.1		1.5 ± 0.1	
ρ^2 [g cm^{-3}]	0.4 ± 0.1		0.4 ± 0.1	

Concluding, hydrophilic PVME chains were not able to dewet from the hydrophobic EtOH-BP1/PS phase due to their high degree of grafting. Thus, in agreement with SPM results, no phase separation process is observed for densely grafted EtOH-BP1 films.

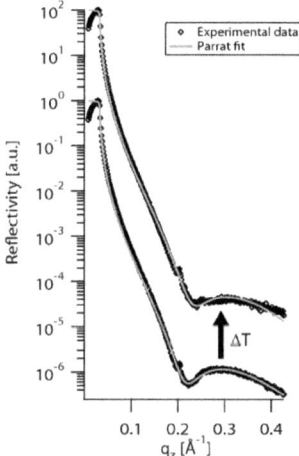

Figure IV.12: X-ray reflectivity curves of the non phase separating EtOH-BP1/PS/PVME blend at RT (lower graph) and 150°C (upper graph); reflectivity curve for the heat treated sample is shifted by a factor of 100.

Thermal response of surface grafted two-dimensional
PS/PVME blend films
-
Effect of grafting densities

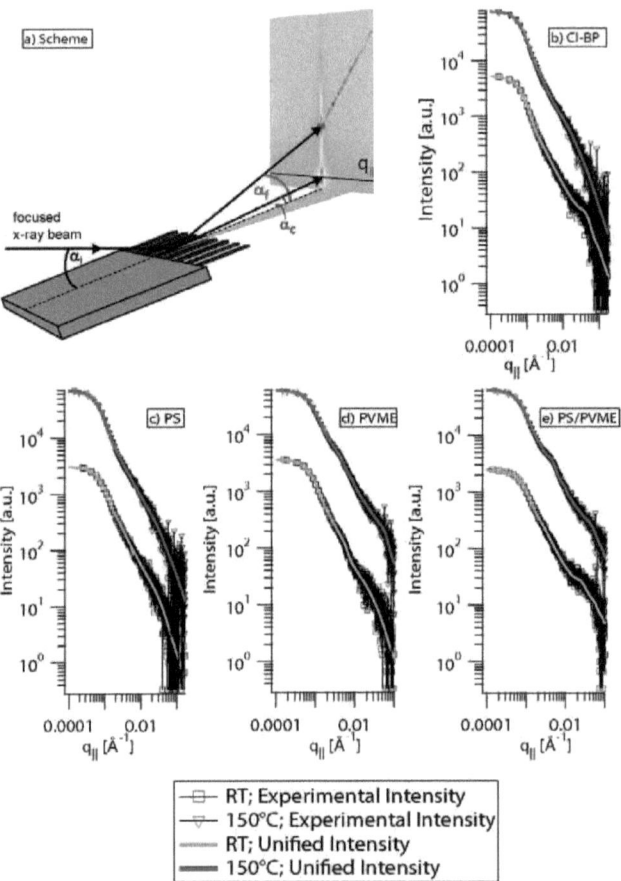

Figure IV.13: µ-GISAXS $q_\parallel$ detector scans of Cl-BP MC sensor systems at RT and 150°C. Black data points represent experimental data, red and brown lines represent the applied fit; a) schematic experimental scattering geometry b) Cl-BP functionalized MC sensor; c) Cl-BP/PS grafted MC sensor; d) Cl-BP/PVME grafted MC sensor; e) Cl-BP/PS/PVME grafted MC sensor; $q_\parallel$ graphs for heat treated specimen are shifted by a factor of 10 for improved visualization.

Congruent lateral film information was obtained from µ-GISAXS data for Cl-BP grafted PS/PVME blend and homo polymer films. Hence it was possible to quantify more accurately domain formations observed during SPM investigations.

Thermal response of surface grafted two-dimensional
PS/PVME blend films
-
Effect of grafting densities

The elongation of the beam-path under grazing incidence geometry allowed to record scattering averaged over the whole coated NCS (Figure IV.13). Detector scans parallel to the surface at the Yoneda peaks maximum were performed to obtain lateral film information in the reciprocal $q_\parallel$ scattering plane (Figure IV.13a). Integrated $q_\parallel$ detector scans were analyzed with the unified fit approach[24, 25, 122]. The applicability of the unified fit approach towards particle systems was thoroughly discussed in chapter III. It was also discussed that power law intensity decays are measures for polymeric mass fractals. The pictured unified fit approach is capable of describing radii of gyrations and average domain centre to centre distances for arbitrary surface and mass fractals. At this point its application was expadended for the analysis of GISAXS from polymeric massfractals. Following this argumentation, R_g values and domain centre to centre distances ξ for polymeric domains with deviations smaller than 20% were obtained.

Scattering data from Cl-BP/PS/PVME films shows that during heat treatment domains with average lateral $R_g = 20$ nm and average centre to centre distances $\xi = 130$ nm are formed. These domains can be related to the polydispers domains observed in SPM images. Such observation is again a clear indication for a phase separation process at RT < T_{LCST} < 150°C. GISAXS from the Cl-BP/PVME film reveals an interference maximum at $q_\parallel \sim 0.002$ Å^{-1} at RT, which shifted to slightly higher $q_\parallel$ - values during heat treatment (Figure IV.13d). From detailed analysis $R_g^{RT} = 30$ nm with a $\xi^{RT} = 102$ nm and $R_g^{150°C} = 21$ nm with a $\xi^{150°C} = 136$ nm can be assigned. In conjunction with µ-XR data the µ-GISAXS experiments prove that PVME domains already dewet at RT from the hydrophobic Cl-BP surface. Annealing leads to shrinking of the PVME domains, caused by contraction of the polymer chains. Rising ξ values can result from a general film collapse of the Cl-BP multilayer or from a second order phase transition in the Cl-BP layer as proposed in chapter IV.4.6. As one can see, fitting parameters for annealed Cl-BP/PVME and Cl-BP/PS/PVME films are equal. Thus, the dewetting of the hydrophilic PVME from the hydrophobic Cl-BP could be raised to T_{LCST} > RT by adding PS. The PS, mixes without dewetting with the Cl-BP (Figure IV.13c). Adding PS to the Cl-silane/PVME system suppresses PVME dewetting at T < RT. However, a PS/PVME enriched second layer of lower density is built.

Thermal response of surface grafted two-dimensional
PS/PVME blend films
-
Effect of grafting densities

IV.4.4 Surface stress results

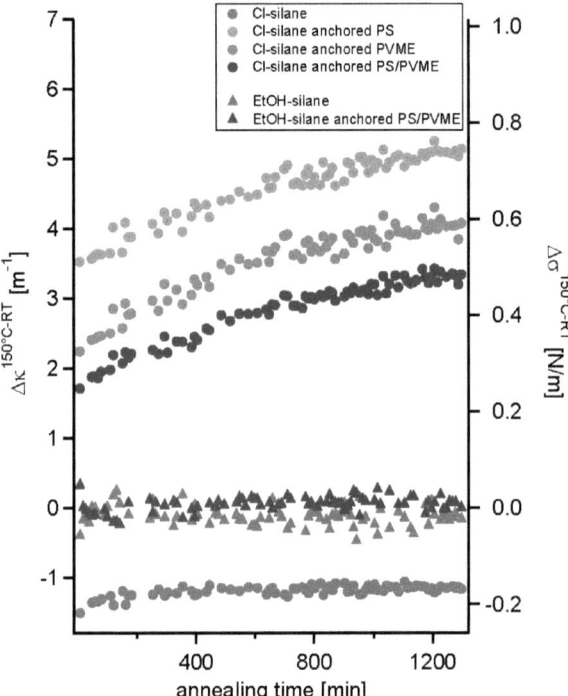

Figure IV.14: Curvature and stress changes vs. time data extracted from interferometric measurements of Cl-BP and EtOH-BP1 systems. All curves were base line corrected for RT values. Curvature and stress data from polymer grafted systems were corrected for reference silane data.

In addition to averaged structural information of the functional Cl-BP/PS/PVME films attractive and repulsive interactions have to be considered in order to complete the obtained physical picture of the phase separation mechanism.

In time dependent annealing experiments surface stresses, which result from transitions in the PS/PVME blend, such as PVME dewetting and influences of polymer/BP grafting points were examined. Polymer uncoated Cl-BP and EtOH-BP1 curvature data taken at 150°C was base line corrected for RT curvature values according to $\Delta\kappa(BP)^{150°C-RT} = \kappa(BP)^{150°C} - \kappa(BP)^{RT}$. Polymer coated MC sensor bending data taken at 150°C was corrected for pure silane curvatures and base

Thermal response of surface grafted two-dimensional PS/PVME blend films
-
Effect of grafting densities

line corrected for RT curvature values according to $\varDelta\kappa(\text{polymer})^{150°C-RT} = \varDelta\kappa(\text{BP/polymer})^{150°C} - \varDelta\kappa(\text{BP/polymer})^{RT} - \varDelta\kappa(\text{BP})^{150°C-RT}$. From $\varDelta\kappa^{150°C-RT}$ data, $\varDelta\sigma^{150°C-RT}$ data was obtained according to Stoney's formula. In such a way conclusions on surface stress changes of the polymer coatings, corrected for effects from transitions in the silane layer can be made.

Starting the discussion from bending data obtained from EtOH-BP1 and EtOH-BP1/PS/PVME films no essential curvature and stress changes can be observed (triangles in Figure IV.14). Such behavior is expected since SPM and XR results clearly indicate that no phase transitions occurred in the highly entropic constrained EtOH-BP/PS/PVME systems.

In contrast pronounced surface stress changes are expected for polymer systems grafted to the low constraining Cl-BP functionalized surfaces. For ungrafted Cl-BP films compressive surface stress changes of $\Delta\sigma^{150°C-RT} = -0.22$ N/m were observed. In contrast, clean uncoated MC sensors of equal thicknesses show tensile stresses due to bimaterial effects in the order of $\Delta\sigma^{150°C-RT} = 0.44$ N/m. Such observed compressive stresses can be related to a partial collapse of the Cl-BP multilayer structure. Compressive surface stresses in Cl-BP layers are reversed to tensile when stresses resulting from PS, PVME and PS/PVME coatings are studied. The total magnitudes of tensile stresses – which were corrected for the Cl-BP reference - resulting from annealing are $\varDelta\sigma^{150°C-RT}$ (PS) = 0.52 N/m, $\varDelta\sigma^{150°C-RT}$ (PVME) = 0.34 N/m and $\varDelta\sigma^{150°C-RT}$ (PS/PVME) = 0.26 N/m. Unraveling the physical reason for the detected tensile stresses, µ-GISAXS results indicate a shrinking of RT dewetted PVME domains during annealing. Polymer domain shrinking is accompanied with entropically driven chain contractions. PVME domains are only 1.7 ± 0.1 nm thick, as measured with µ-XR. Such small domain thicknesses clearly indicate that the polymer chains are grafted at more than one chain segment to the MC sensor substrate. The resulting attractive force can therefore be transferred to the MC sensor substrate. Consequently tensile stresses are measured. The density of the grafting Cl-BP is equal within one studied MC sensor array. In such a way similar grafting point densities can be assured for all polymer-films grafted to the same MC sensor array. This makes comparisons of the different polymer systems possible. Equal Cl-BP layers result in similar grafting point densities for all Cl-BP films. Consequently, attractive forces within the polymer films resulting from the discussed entropy changes can be assumed to be equal and irrespective of the molecular weight for all three grafted films, according to

Thermal response of surface grafted two-dimensional
PS/PVME blend films
-
Effect of grafting densities

$$\Delta F_{entrop.}^{Cl-silane/PS} \approx \Delta F_{entrop.}^{Cl-silane/PVME} \approx \Delta F_{entrop.}^{Cl-silane/PS/PVME} < 0 \qquad (\text{IV.4.1})$$

Following equal tensile surface stress proportions are expected by only regarding entropic chain contractions. Nevertheless, the total magnitudes of tensile stresses are clearly dependent on the grafted polymer systems. It can be deduced that superimposed repulsive forces, resulting from surface and interfacial energies cannot be neglected. For such argumentation the magnitudes and directions of the forces resulting from surface energy and interfacial energy changes have to be discussed.

First the surface energy of the film/air interface can change upon annealing. Such energy changes can be attributed to changes in surface tensions of the grafted polymers. Bulk PS and PVME have similar surface tensions of $\gamma(PS)^{RT} = 40.2$ mN/m and $\gamma(PVME)^{RT} = 36.0$ mN/m at RT[115]. For both bulk polymers experimental and theoretical approaches[137-139] predict a nearly linear decrease of surface tensions according to $-\dfrac{d\gamma(PS)}{dT} = 0.072$ and $-\dfrac{d\gamma(PVME)}{dT} = 0.075$ with temperature. Such decreases in surface tensions lead to a decrease in the Gibbs surface energy after the relationship $\Delta G_{Surface}^{150°C-RT} \approx -A\Delta\gamma^{150°C-RT}$. Since the surface stress can be directly related with the Gibbs surface energy[40], reductions in Gibbs surface energies in the grafted films lead to compressive stresses of similar magnitude. Obviously the different magnitudes for surface stress reductions cannot be explained exclusively by reductions in surface energies.

Thus, forces resulting from interfacial energy changes have to be regarded in order to explain the variation in tensile stresses detected for grafted homo and blend polymer films. The only interfacial contribution, which was not similar for all systems and which is still reflected in differential $\Delta\kappa^{150°C-RT}$ and $\Delta\sigma^{150°C-RT}$ data, can result from the Cl-BP/polymer interface.

From µ-GISAXS and µ-XR results follows that the Cl-BP/PS film was in a microscopically mixed state at RT and did not phase separate during annealing. Thus, interaction energy changes can be assumed to be small. The Cl-BP/PVME film was already found in a dewetted two phase state at RT as indicated by µ-GISAXS and µ-XR results. From the attractive chain contractions can be argued that intramolecular PVME chain interactions are favored, while intermolecular Cl-BP/PVME interactions are disfavored. In other words, PVME domains repel the Cl-BP grafting layer, which results in a lateral pressure. This conclusion is further supported by the found increase in average

Thermal response of surface grafted two-dimensional
PS/PVME blend films
-
Effect of grafting densities

domain center to center distances ξ from 102 nm to 136 nm as measured from μ-GISAXS. Compared to Cl-BP/PS films, such interfacial repulsive forces led to a MC sensor bending away from the coating layer in the Cl-BP/PVME film. The Cl-BP/PS/PVME is found not entirely mixed at RT as indicated by the found two phase film (Table V.3.1). However, PVME domain formation was not observed at RT (μ-GISAXS and SPM). Annealing caused pronounced PVME domain formation of similar domain sizes and domain centre to centre distances than for the Cl-BP/PVME film as measured with μ-GISAXS. The lower surface stress change detected for the Cl-BP/PS/PVME film compared to the Cl-BP/PVME film can therefore be attributed to an increased repulsion between the PVME chains and the mixed Cl-BP/PS phase. Following repulsive interfacial forces of the three comparable grafted polymer film systems are related according to:

$$0 \approx \Delta F_{interface}^{Cl-silane/PS} < \Delta F_{interface}^{Cl-silane/PVME} < \Delta F_{interface}^{Cl-silane/PS/PVME} \qquad (IV.4.2)$$

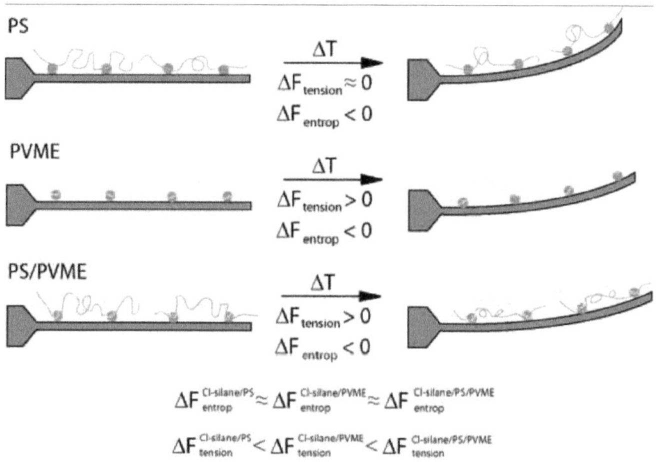

Figure IV.15: Illustrative scheme of the proposed mechanisms resulting from annealing the grafted polymer systems on a molecular level, including NCS bendings and proposed entropic driven attractions and interfacial repulsions.

During constant annealing tensile stresses in all polymer films increase further by 0.2 N/m. These changes are equal for all polymer anchored films. They are therefore not attributed to individual stress changes caused by interfacial interaction energy changes, but to continuous structural and conformational rearrangements. Summarizing it can be concluded that attractive entropic driven conformation changes dominate repulsive surface energy effects and individual repulsions driven by interfacial interaction energy changes. Thus dominating tensile stresses are detected (Figure IV.15).

IV.4.5 Summary

Within this chapter was shown that phase transition processes in grafted two-dimensional PS/PVME polymer films can be tuned with the density of active benzophenone grafting points. The presented results show that hydrophobic benzophenone linking layers with low grafting point densities at the surface are necessary for the observation of PVME dewetting in the studied temperature range. In such a way it was possible to raise the blend's LCST above room temperatures. Such a behavior was not reported for spin cast two-dimensional films, where the LCST was found to be below room temperature[115]. Apart from a raise in LCST experimental results propose similar PVME dewetting mechanisms than proposed in spin cast films. Due to conformational constraints caused by chain grafting at more than one segments, PVME chains are not able to recover all their conformational energy. In combination to dewetting processes a collapse of the formed linker multilayer structure was observed. However, such a film collapse did not disturb the observed phase separation process. Combining structural investigations with surface stress investigations extracted from MC sensor bending experiments allowed transferring results of the films structure towards energetic changes in the polymer films upon phase transition. Detected tensile stresses were related to attractive entropic spring mechanisms during conformational recovering of constrained chains. It was further proposed that attractive entropic effects were lowered by opposing repulsive processes, driven by decreasing surface tensions and interfacial repulsions. However attractive effects dominate repulsive effects during phase transition.

Thermal response of surface grafted two-dimensional
PS/PVME blend films
-
Effect of grafting densities

IV.4.6 Grafting densities in fully active BP films

Before systematically varying the BP grafting density by e.g. fractional reduction of the carbonylic unit by different concentrations of sufficient reducing agents (not performed in this thesis), grafting density effects were studied for fully active BP molecules. Such studies were performed using the EtOH-BP compound, since its reactivity towards the SiO_x surface can be tuned with different solvents. Using the EtOH-BP1 preparation route with EtOH acting as solvent, equilibrium between the solvated EtOH-BP and the surface bound EtOH-BP is assumed (Figure IV.16). In contrast the equilibrium is shifted towards the product side using toluene as solvent (EtOH-BP2).

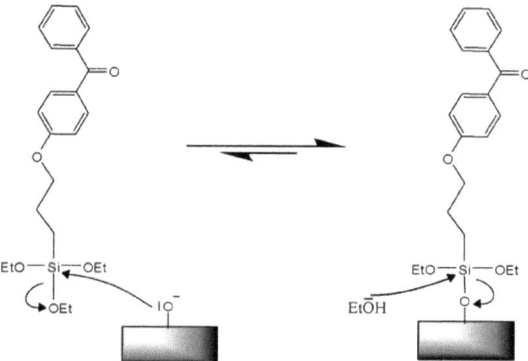

Figure IV.16: Mechanism of EtOH-BP1 coating procedure.

Thus EtOH-BP2 films are assumed to be of higher density than EtOH-BP1 films. Contact angle experiments with deionized water showed contact angles of 62 ± 2° for EtOH-BP1 systems, while contact angles of 67 ± 2° were found for ETOH-BP2 systems. Due to intrinsically high errors in contact angle experiments, the statement of higher surface coverage of the ETOH-BP2 compared to EtOH-BP1 cannot be properly verified. For comparative experimental verification of the difference in surface coverage x-ray reflectivity experiments were carried out for EtOH-BP1 and EtOH-BP2 systems. Each sample set consisted of four samples. One Si-specimen was coated only with the silane. The other three were grafted with the PS and PVME homopolymer, and with the PS/PVME = 20/80 blend.

Thermal response of surface grafted two-dimensional PS/PVME blend films
-
Effect of grafting densities

Reflectivity scans shown in Figure IV.17 show that Fresnel oscillations move towards lower q_z for all EtOH-BP2 films. Thus, thicker films were formed within the EtOH-BP2 preparation route. Thicker film formations within the pure silane films indicate that EtOH-BP2 molecules are dense packed and more stretched away from the surface in contrast to the EtOH-BP1 preparation route.

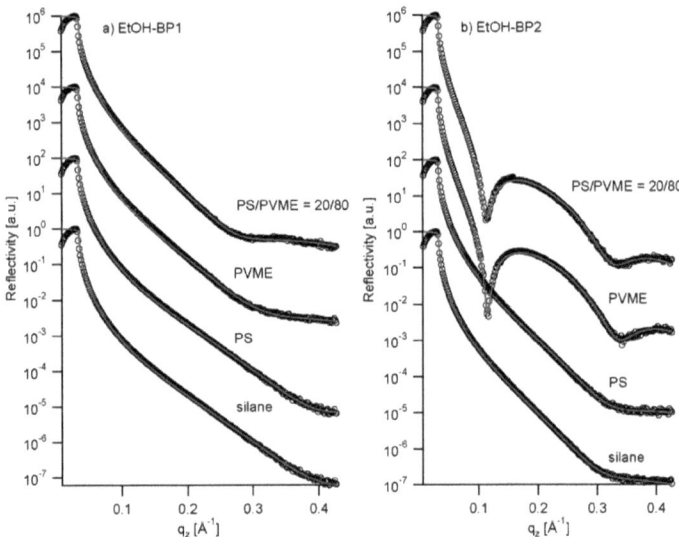

Figure IV.17: X-ray reflectivity scans for a) EtOH-BP1 and b) EtOH-BP2 samples. Black data points represent experimental reflectivity; red line represents calculated reflectivity according to tanh expanded step models; for clarity the curves are shifted by a factor of 100 against each other.

For both, the EtOH-BP1 and EtOH-BP2 prepared pure silane and PS films are of equal film thickness, while PVME and PS/PVME are thicker by an equal magnitude prepared films. This result is consistent with results from Cl-BP films (Figure IV.11, Table IV.4.1).

Thermal response of surface grafted two-dimensional
PS/PVME blend films
-
Effect of grafting densities

Table IV.4.2: film thicknesses obtained from analysis of reflectivity profiles for EtOH-BP1 and EtOH-BP2 systems, presented in Figure IV.17

	t (EtOH-BP1) [Å]	t (EtOH-BP2) [Å]
silane	8 ± 1	11 ± 1
PS	8 ± 1	10 ± 1
PVME	10 ± 1	28 ± 1
PS/PVME	11 ± 1	29 ± 1

Interestingly the higher surface coverage obtained with the EtOH-BP2 preparation route led to an increase in film thickness by 18 Å. In contrast PVME and PS/PVME grafting led only to thickness increases of 2 Å for EtOH-BP1 prepared films. This result is probably caused by two cooperative effects. First, the dense EtOH-BP2 linking layer is able to graft higher amounts of polymer material than the EtOH-BP1 linking layer. Second, higher surface coverage leads to higher interfacial EtOH-BP/polymer energies. Thus the grafted polymer is repelled from the surface, which leads to the high film thicknesses.

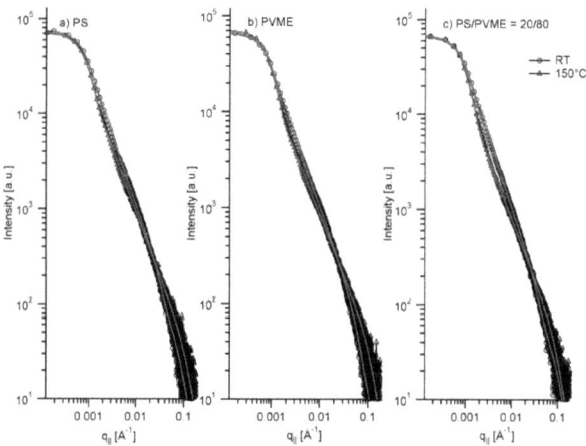

Figure IV.18: : µ-GISAXS $q_\parallel$ detector scans of EtOH-BP2 MC sensor systems at RT and 150°C. Black empty circles represent GISAXS at RT; empty triangles represent GISAXS at 150°C a) PS grafted MC sensor; b) PVME grafted MC sensor; c) PS/PVME grafted MC sensor; red lines are represent unified calculated GISAXS.

Thermal response of surface grafted two-dimensional
PS/PVME blend films
-
Effect of grafting densities

Due to the high film thickness, one could expect that the polymer loops within the 18 Å film layer have enough degrees of freedom to allow PS/PVME phase separation. This possibility was studied with µ-GISAXS (Figure IV.18). Similar to studies for the Cl-BP/PS/PVME systems, the polymer was grafted to a prefunctionalized MC sensor array. Thus, EtOH-BP2 layers of equal surface coverage were obtained. However, µ-GISAXS from the PS/PVME blend grafted film did not show any correlation peak at 150°C. Therefore, the EtOH-BP2 grafted PS/PVME blend did not phase separate within the studied temperature range. In contrast to µ-GISAXS obtained from Cl-BP grafted PVME films, no correlation peak was found at RT and 150°C. It can be concluded that high grafting densities lead to uniform PVME enriched layers (~ 18 Å) on top of the EtOH-BP2 layers as proposed from XR. The high grafting density does not allow the free polymer loops within the PVME enriched layer to dewet from the EtOH-BP2 and EtOH-BP2/PS enriched layer.

Subsequent the influence on surface stresses of higher EtOH-BP2 surface coverage in contrast to EtOH-BP1 films was studied. Similar to the coating procedure for Cl-BP films, two MC sensors of one array were grafted with the PS, PVME homopolymers and with the blend, respectively. Two reference MC sensors (only EtOH-BP2) were left uncoated of polymer. Different to surface stress experiments conducted for Cl-BP and EtOH-BP1 film systems, the temperature was increased by 2°C/min, while taking one topography image per 30 s. Thus, the detection of possible phase transition temperatures became possible (Figure IV.19). Surface stress data was recorded for a heating ramp from 20 - 150°C and for a subsequent cooling ramp from 150°C – 20°C.

Differential curvature data of polymer grafted MC sensors were obtained according to $\Delta\kappa$ (polymer) = κ (polymer) – κ (EtOH-BP2). Curvature data for ungrafted EtOH-BP2 MC sensors was taken as obtained from data analysis. Accordingly, surface stress and differential surface stress data were calculated with Stoney's formula (Eq. (II.1.3)). Surface stress data of EtOH-BP2 shows that initial tensile stresses of $\sigma = 0.1$ N/m increased with $\frac{d(\sigma(EtOH-BP2))}{dT} = 1.8 \cdot 10^{-3} \frac{N}{m°C}$ towards 1.8 N/m. This increase is probably caused by a bimaterial effect. At T ~ 80°C tensile stresses turn to compressive with $\frac{d(\sigma(EtOH-BP2))}{dT} = -4.0 \cdot 10^{-3} \frac{N}{m°C}$. Such surface stress changes are possibly attributed to phase transitions in the EtOH-BP2 layer. A possible reason is the transition from a rigid EtOH-BP2 layer towards a softer layer, where benzophenone groups become more flexible.

Thermal response of surface grafted two-dimensional PS/PVME blend films - Effect of grafting densities

For cooling a similar curve progression than for heating is observed. However, the observed curve offset indicates that surface stresses are not totally reversible. Differential surface stresses for all polymer grafted MC sensors are constant from 20°C < T < 80°C. Thus bimaterial effects observed for EtOH-BP2 coated MC sensors were similar for polymer grafted MC sensors.

For T > 80°C differential tensile stress changes of $\frac{d(\Delta\sigma(PS))}{dT} = 3.8 \cdot 10^{-3} \frac{N}{m°C}$, $\frac{d(\Delta\sigma(PVME))}{dT} = 6.0 \cdot 10^{-3} \frac{N}{m°C}$ and $\frac{d(\Delta\sigma(PS/PVME))}{dT} = 3.6 \cdot 10^{-3} \frac{N}{m°C}$ were measured. It may be concluded that the grafted polymer chains inhibit the flexibility of the EtOH-BP2. Consequently the films E-moduli are not decreasing and $\frac{d(\Delta\sigma(polymer))}{dT} + \frac{d(\sigma(EtOH-BP))}{dT} \geq 0$ is measured for T > 80°C. Similar to ungrafted EtOH-BP2 films differential surface stress progression for cooling was found to be similar except for an offset at T = 80°C. In addition to these experiments, reference surface stress experiments on ungrafted polymer films of similar thickness have to be performed in future studies in order to check the effect of thermal expansion by the polymer itself.

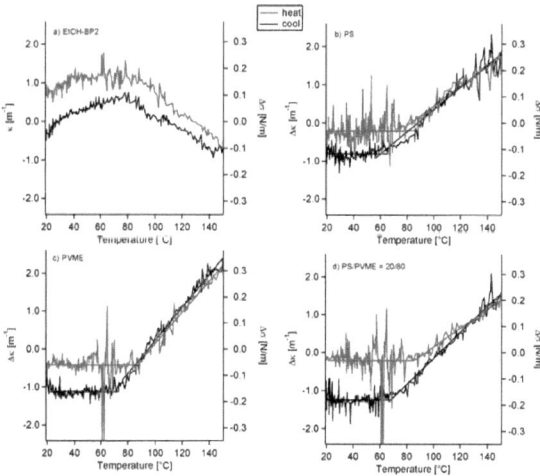

Figure IV.19: Curvature and differential stress changes vs. temperature data extracted from interferometric measurements of EtOH-BP2 systems.

Thermal response of surface grafted two-dimensional PS/PVME blend films
-
Effect of grafting densities

The hypothesis of a phase transition of the EtOH-BP2 film was supported by first comparative DSC measurements for the dried EtOH-BP bulk sample (Figure IV.20). Two transitions can be observed from the DSC thermogram. At -67°C, an endothermal second order transition is observed. It can be concluded that the EtOH-BP is in an amorphous glassy state at T < -67°C and turns into a viscous fluid for T > -67°C. A second phase transition is observed at T = 88°C. This second order phase transition is of exothermal nature. A possible reason for this phase transition is a decrease in viscosity of the oily fluid, observed at RT. Such transitions would support the predicted enhanced EtOH-BP flexibility at T > 80°C from surface stress experiments. However, the dynamic properties of the EtOH-BP upon this phase transition have to be studied in detail with e.g. rheology and NMR in order to fully support the used hypothesis.

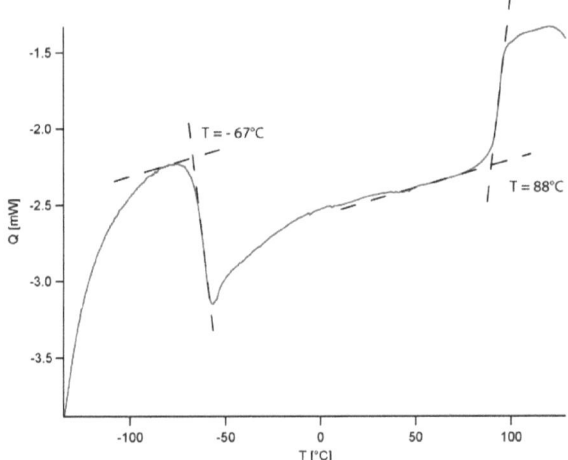

Figure IV.20: DSC thermogram of EtOH-BP

At T > 130°C the EtOH-BP started to degrade. However, no indication for degradation was observed in prepared films as observed from surface stress experiments. The degradation is therefore attributed to Si-O-Et bond disjunctions in the bulk material.

V Stress/structure correlation in grafted from PMMA brushes

V.1 Motivation

Based on the work of Bumbu et al.[58] this chapter addresses experimental results on the collapse/swelling transition of PMMA brushes. In the previous work PMMA brush coated MC sensor arrays were used to investigate surface stress changes upon different solvent qualities. The solving quality was tuned by mixtures of a good and a bad solvent. The previous work unraveled that the coated MC sensors bend away from the coating after increasing the amount of good solvent. However, MC sensor bending experiments were limited to relative deflection methods and no conclusions on the initial surface stresses in the collapsed brush state (bad solvent conditions) could be drawn. It was therefore not possible to distinguish between initial non bent MC sensors, which bend downwards, and initial upwards bent MC sensors, which relax towards smaller curvatures. In the first case obtained bending data could be explained with an increase in lateral surface pressures resulting from solvent adsorption into the brush. In the second case bending data could be explained by a surface stress decrease caused by initial attractive polymer-polymer interactions, which relax in the good solvent regime. Such differentiations can now be drawn with the used NIR imaging interferometric technique, by directly recording 3-D topographies of the MC sensor arrays. In addition, it was observed that the surface stress progression upon the collapse/swelling transition was highly dependent on collapse/swelling route. It was suggested that non negligible kinetic effects were observed during collapse/swelling transitions.

Bumbu et al. compared the obtained stress data with the theoretical approach from Birshtein and Lyatskaya.[30] (Chapter II.3.2.2). χ-parameters used for estimations of brush-thicknesses Eqn. ((II.3.17),(II.3.18)), were obtained from solubility parameter estimations developed from Hildebrand and Scott[139]. Such parameter estimations are based on bulk solution theories. Applications of such theories neglect entropy constraints in densely grafted polymer brushes and wetting effects. From NR experiments the molar fraction of deuterated solvent in respect to hydrogenated polymer can be obtained. After knowing the total amount of solvent adsorbed in the brush phase for several solvent bulk mixtures, fitting curves can be simulated from Eqn. ((II.3.17),(II.3.18)), and reasonable solvent-brush χ-parameters can be obtained. Comparing effective χ-parameters from Eq. (II.3.19) with surface stress data allowed adequate understanding of the transition from the collapsed towards the swollen PMMA brush.

V.2 Grafting from prepared polymer brushes

V.2.1 Introduction to grafting from with atomic transfer radical polymerization (ATRP)

Densely endgrafted polymer brush films of thicknesses bigger than the bulks R_g can be prepared by **"grafting from"** techniques. Here a polymerization initiator is irreversibly bound to the surface. Such can be done either by thiol/gold or silane/SiO_x chemistry. The polymerization starts directly at the surface. Since surface initiators can be usually densely packed, polymer brushes with high grafting densities, which are vertically stretched away from the surface, can be obtained. Owing to their high grafting densities and high stretching behavior, such grafted polymer films are commonly called polymer brushes. Polymerization can be performed on the basis of standard bulk polymer synthesis techniques[132, 140], such as anionic, carbocationic, conventional radical or controlled radical polymerization techniques. Especially radical grafting from techniques are found in many grafting from applications, due to lower reactivity of radicalic chain ends towards impurities such as traces of water, oxygen or compared to anionic chain ends. Radical grafting from techniques are not limited to specific solvents and can therefore also performed in aqueous solvent environments. However, recombination and disproportionation reactions lead to distributions in molecular weights. Such would make descriptions of reflectivity profiles more complex, since polydisperse brushes can loose their stepfunction/parabolic density profile[141]. Therefore the use of a controlled radical polymerization technique is required, in order to control molecular weights and obtain low molecular mass distributions. Such controlled "living" radical polymerizations are e.g. "Nitroxide Mediated Polymerization"[142], "Reversible Addition Fragmentation Chain Transfer" polymerizations (RAFT)[143, 144] or the within this work used "Atomic Transfer Radical Polymerization" (ATRP)[28, 29, 145].

Within common ATRP routes polymerization starts from halogenated (X) initiators. They are reduced to their radical form with the use of complexes, composed of a transition metal (M^n) in oxidation state n and chelating ligand (L). The complex (M^n/L) turns into its oxidized state X-M^{n+1}/L. The activated initiator reacts with the monomer (Figure V.1). In contrast to traditional radical polymerization techniques, the mechanism is reversible with certain rate constants k_{act}, k_{deact} and k_p, for the activation, deactivation and polymerization step. These rate constants can be properly adjusted by thorough adjustment of the catalytic component and reaction media. Small ratios of

Grafting from prepared polymer brushes

k_{act}/k_{deact} lead to small amounts of the activated species $P_n\dot{}$. Consequently termination reactions are suppressed and narrow distributed polymer weights are obtained.

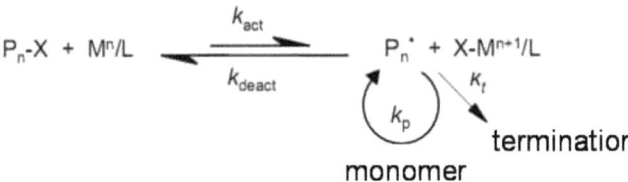

Figure V.1: Reaction scheme of ATRP

For surface initiated polymerization the ATRP initiator is immobilized to a substrate, by e.g. introduction of silane functionalizations. Compared to solution ATRP the number of active Pn-X groups would be dramatically reduced, because they are only located at the substrates surface. Thus, free (non surface bound) initiator has to be added to the ATRP reaction mixture in order to obtain comparable kinetics than in solution ATRP. Such procedure has the advantage that the obtained free polymer can be used for polymeric weight determinations, assuming similar polymeric weights for free and surface bound polymers[146]. A second possibility is the addition of deactivating $X-M^{n+1}/L$ species, which actively deactivates the radical $P_n\dot{}$ form, to the reaction mixture[28].

V.2.2 PMMA brushes prepared with surface initiated ATRP

The synthesis of densely endgrafted PMMA brushes on Si MC sensor array substrates and Si – wafer substrates was performed as illustrated in Figure V.2 [147]. The surface initiator (4) was synthesized starting from an esterification of 2-bromo-2-methylpropionylbromide (2) with allylalcohol (1). The obtained allyl-compound was further processes by hydrosilation with dimethylchlorosilane and the initiator (3) was obtained. The initiator was immobilized to the precleaned Si – substrates. Si - MC array and Si – wafer substrates were cleaned with standard base cleaning procedures in order to obtain controlled hydration of the native SiO_x surface (chapter II.5.1).

The surface initiator is immobilized to the substrates via the active Cl-silane functionalization. The employed base triethylamine acts as a scavenger for generated protons, favoring the immobilization

Stress/structure correlation in grafted from PMMA brushes

Grafting from prepared polymer brushes

reaction. The ATRP is carried out employing Cu(I)Br with the chelating N,N,N`,N`,N``-pentamethyldiethylenetriamine (PMDETA) ligand as ATRP catalyst. The sacrificial ATRP solution initiator Ethyl 2-bromoisobutyrate was added to control the reaction. The free initiator creates necessary concentrations of Cu(II) complexes, which controls the polymerization in solution and initiated from the substrate. Anisole was chosen as solvent, because of its good solving properties for the employed Cu complexes. After completion of polymerization the coated samples have to be thoroughly extracted in order to remove all adsorbed and not chemically bound PMMA from the surface. Fractions of the free PMMA were precipitated from MeOH for molecular mass determinations by Gel Permeation Chromatography (GPC).

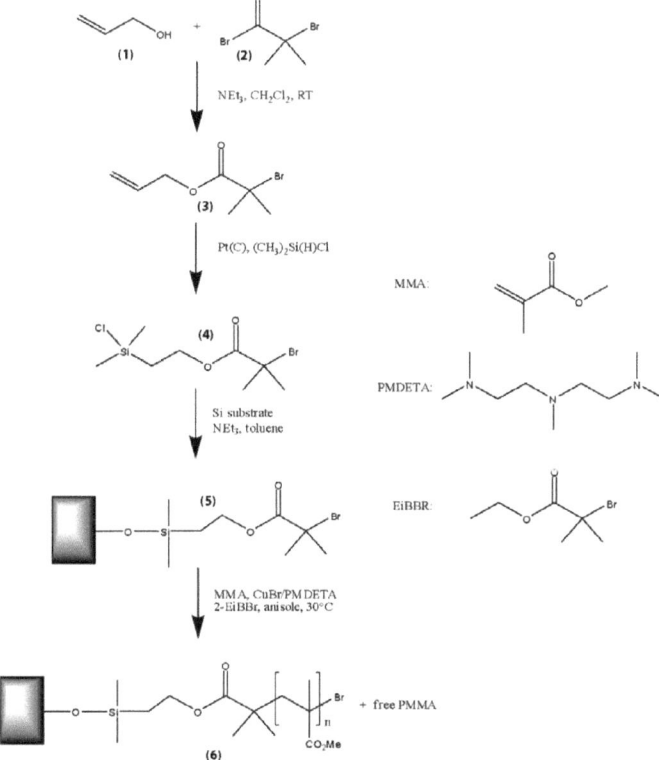

Figure V.2: Reaction Scheme of grafting from PMMA brush synthesis

V.2.3 Simultaneous MC sensor array/wafer coating

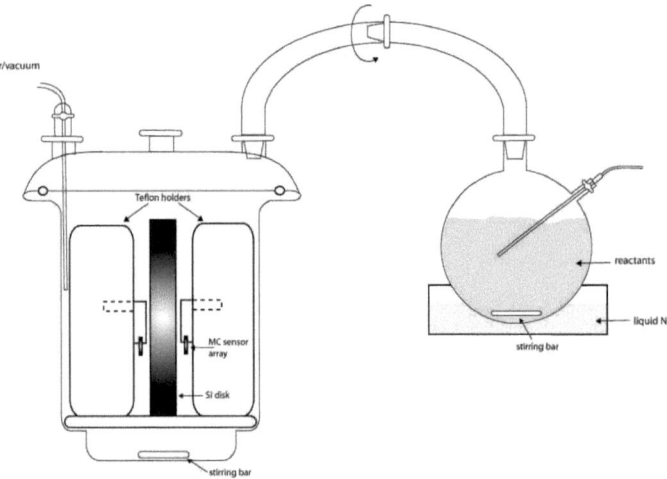

Figure V.3: Reaction apparatus for simultaneous PMMA ATRP brush synthesis on MC sensor arrays and Si-wafer disks. The substrates were located in custom made Teflon holders in a plane flange beaker (left hand side). The reactor could be closed with a ceiling O-ring. The reaction mixture was provided in a 1L Schlenk flask (right hand side). The degassed reaction mixture could be added under inert atmosphere into the plane flange reactor by rotating the curved glass tubes.

For reliable comparisons of results obtained from NR measurements and surface stress investigations, equal brushes have to be synthesized for both kinds of samples. Since brush thickness and grafting density depends on the molecular weight of the polymer and the grafting density of the surface initiator, the best choice is to perform all the necessary reaction steps for both kinds of samples simultaneous in one reactor. MC sensor arrays were typically 2.5 mm wide, 3.5 mm long and 0.5 mm thick. The used disk shaped Si substrates used for NR measurements were 100 mm in diameter and 15 mm thick. Since Si disks for NR measurements were by a factor of ~ 100 bigger than MC sensor arrays, a specialized reactor for simultaneous sample treatment was designed (Figure V.3).

Stress/structure correlation in grafted from PMMA brushes
-
Grafting from prepared polymer brushes

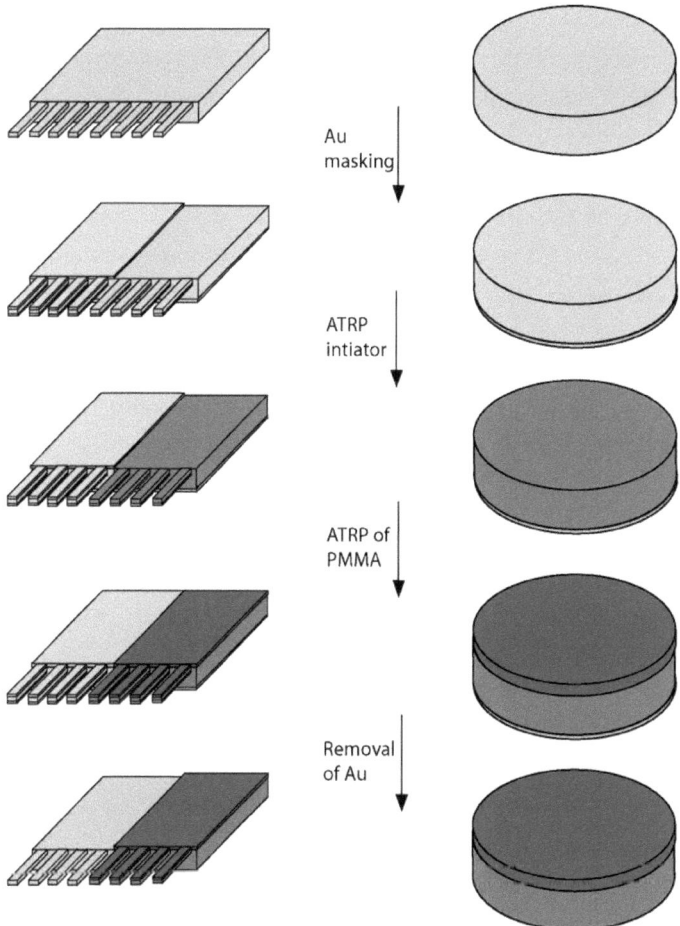

Figure V.4: Simultaneous processing scheme for MC sensor array and Si-disk specimen, starting from gold protected MC arrays and Si-disk.

After cleaning four of the MC sensors topsides were coated with the PMMA brush. Therefore the MC sensor arrays backside and half of the arrays topside was masked passivating Au film as illustrated in Figure V.4 and described in chapter II.5.2.

For the sake of equal processing a 30 nm protecting gold layer was sputtered on one of the Si-disk sides (First step in Figure V.4).

Stress/structure correlation in grafted from PMMA brushes - Grafting from prepared polymer brushes

Following the surface ATRP initiator was applied simultaneously to the specimen (Second step in Figure V.4). For that purpose the reaction mixture of 400 mL water free toluene, 3.3 mL NEt_3 and 6.4 mL as prepared initiator was provided in a clean 1L Schlenk flask under Ar atmosphere. Following the prepared mixture was transferred under Ar atmosphere to the plan flange reactor as illustrated in Figure V.3. Both kind of specimen were totally covered by the reacting mixture. The reaction was stirred for 12 h at room temperature. After silanization MC sensor arrays were cleaned with CH_2Cl_2 for 2 h using soxhlet extraction, while Si-disks were continuously rinsed with CH_2Cl_2.

For ATRP PMMA brush synthesis the specimen were located in the freshly cleaned plane flange reactor (Third step in Figure V.4). The whole reaction apparatus was assembled as illustrated in Figure V.3 and properly ceiled. The whole apparatus was set under Ar atmosphere and the reaction mixture of 200 mL MMA, 300 mL, 513 mg CuBr and 750 µL PMDETA was added to the illustrated Schlenk flask. As last reactant 530 µL of the free initiator EiBBr was added and the Schlenk flask was immediately frozen with liquid nitrogen to prevent polymerization. The apparatus was set under vacuum and the frozen mixture was thawed and degassed. The freezing/thawing cycle was repeated three times until most of the solvated gas was removed. Following the cold polymerization mixture was transferred into the plan flange reactor until both kind of specimen were totally covered by the reaction mixture. The ATRP was carried on under stirring and continuous passing of Ar for 40 h at 30°C.

After polymerization the PMMA coated specimen were thoroughly cleaned from adsorbed and non bound free PMMA with CH_2Cl_2. Fractions of the free PMMA from solution polymerization was precipitated under drop wise addition of the solution to an excess of MeOH. For the purification, the filtered precipitate was solved again in THF. The precipitation process was repeated until a white precipitate was obtained. The obtained PMMA was analyzed with GPC. For the studied sample M_n = 23400 g/mol with a PDI of 1.2 was obtained from GPC analysis.

As a last step the protecting gold layers were removed with KI/I_2 solution in MilliQ water. The specimen were continuously immersed in the KI/I_2 solution and rinsed with MilliQ water in order to remove all gold residuals. The dried films were 7.1 ± 0.05 nm thick as analyzed with x-ray reflectomery from Si-disk specimen. The macroscopic density was calculated to be $\rho = 1.0 \pm 0.05$ g/cm³.

Stress/structure correlation in grafted from PMMA brushes
-
Grafting from prepared polymer brushes

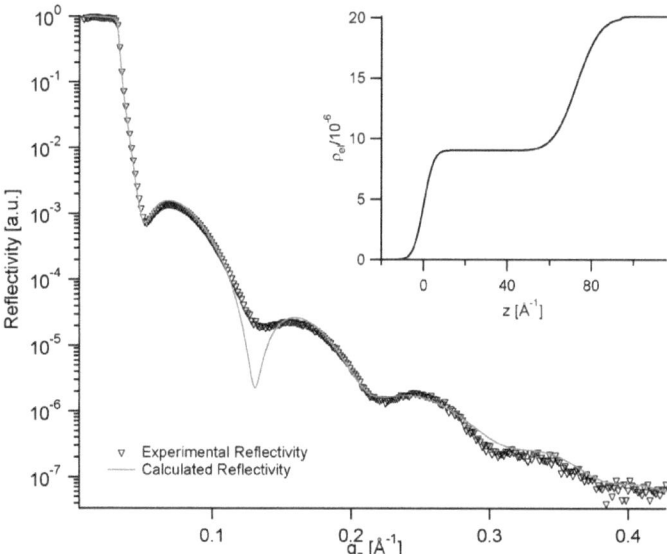

Figure V.5: X-ray reflectivity profile of dry PMMA brush film as measured from Si-disk specimen. From Step function modeling the electron density profile was obtained (inset); $z = 0$ position indicates the PMMA/air interface and $z = 7.2$ nm indicates the Si/PMMA interface.

V.3 Neutron reflectivity results

V.3.1 Experimental objective

Within this chapter the collapse/swelling transition and its reversibility of the prepared PMMA brush was studied with neutron reflectivity.

It was previously observed that chain entanglements in a dry thin PMMA film are generally preserved by changing the film geometry to a densely grafted polymer brush[148]. The question arises, if chain entanglements in dry polymer brushes are preserved during solvent exposure. Since chain entanglements are kinetically stabilized a subsequent question is, if the observed entanglement effects are dependent on the collapse/swelling route. Are kinetic effects present?

Consequently structural brush attributes such as brush height and solvent uptake should be different, within predominant kinetics. These possible phenomena were studied by neutron reflectivity. The dried brush was first immersed in bad solving MeOH (latter in the text: collapsed brush of dry origin). Solvent exchange of MeOH with good solving THF unraveled structural changes during fast swelling processes. Subsequently and during the same experimental beam time the swollen brush was collapsed again (collapsed brush of swollen origin). Consequently changes during fast collapse processes could be measured and conclusions on kinetic entanglement effects could be drawn. Still during the same experimental series, the bulks volume fraction of good solving THF was gradually increased within several days. From such obtained data χ-parameter sets were obtained using the pictured theory from Birshstein and Lyatskaya[30] and allowed a comparison of χ-parameters predicted from bulk theory. Determination of χ-parameter values allowed to discuss the MeOH desorption/THF adsorption behavior.

V.3.2 Data treatment

For studying collapse swelling mechanisms of the ATRP synthesized PMMA brushes structural *in situ* features of the brushes have to be characterized. *In situ* denotes that the polymer brush is immersed in the particular good or bad solvent mixture. Solvent mixtures were prepared from mixtures of bad solving deuterated methanol (d4-MeOH; denoted as THF throughout this chapter)

Stress/structure correlation in grafted from PMMA brushes - Neutron reflectivity results

and good solving deuterated tetrahydrofuran (d8-THF; denoted as THF throughout this chapter). As discussed in chapter II.2.2 neutron reflectivity is an expensive but powerful method to characterize structural properties of polymer films in deuterated solvents. Hence, statements on PMMA brush thicknesses, estimations on surface roughnesses and refractive indices, which can be related to the volume fraction of solvent in the brush layer, can be made. Experimental reflectivity data was corrected for the background signal including off specular oscillations, which can result from reflectivity at correlated brush interfaces as described in chapter II.2.4.

From fitting complete parameter sets for the brush height H, scattering length density (SLD) for the brush phase and statistical roughness s could be obtained. The SLD^{bulk} parameter was constrained for calculated $SLDs$ for each bulk solvent composition with the use of a free accessible online scattering length density calculator[149]. All reflectivity profiles were fitted with a *tanh* expanded Alexander - de Gennes (AdG) brush model[26, 27]. It was shown in chapter II.2.3 and chapter II.3.2.1 that roughness profiles of swollen polymer brushes are best described with a parabolic MWC model. However, the free energy and resulting brush height of both model types (AdG and MWC) scale with N^1 and are directly proportional. Moreover, for the studied comparably thin brush films only two oscillatory minima were observed in the experimental q-range. For that reason, fitting qualities did not increase using the MWC model instead of the AdG model. A third reason for the use of the AdG model is that the theory from Birshtein and Lyatskaya[30], which describes the collapse/swelling transition used for χ-parameter estimation is based on the AdG model. Moreover, the use of the AdG model for describing polymer brushes is also supported by recent theory[150].

The SLD is related to the sum of the neutron's scattering lengths b_i and the molar volume v_{m_i} for each compound i according to

$$SLD = \frac{\sum b_i}{\sum v_{m_i}} \quad (V.3.1)$$

Thus, the obtained SLD^{brush} is a sum of SLD^{brush} (MeOH + THF) and SLD^{brush} (PMMA). Accordingly, the volume fraction ϕ^{brush} (MeOH + THF) is related to SLD^{brush} and SLD^{bulk} by

$$\phi^{brush}(MeOH+THF) = 1 - \frac{SLD^{brush} - SLD^{bulk}}{SLD^{theo}(PMMA) - SLD^{bulk}} \quad (V.3.2)$$

Neutron reflectivity results

In order to obtain integrated ϕ^{brush} values for the whole brush layer the obtained density profiles were integrated and normalized to the brush height H according to

$$\phi_{tot}^{brush}(MeOH+THF) = \frac{\int_0^H \phi^{brush}(MeOH+THF)dz}{H} \qquad (V.3.3)$$

V.3.3 Fast collapse/swelling process

As mentioned above, possible kinetic entanglement effects during the fast swelling process were investigated. Figure V.6 shows the reflectivity profile of the MeOH immersed PMMA brush specimen.

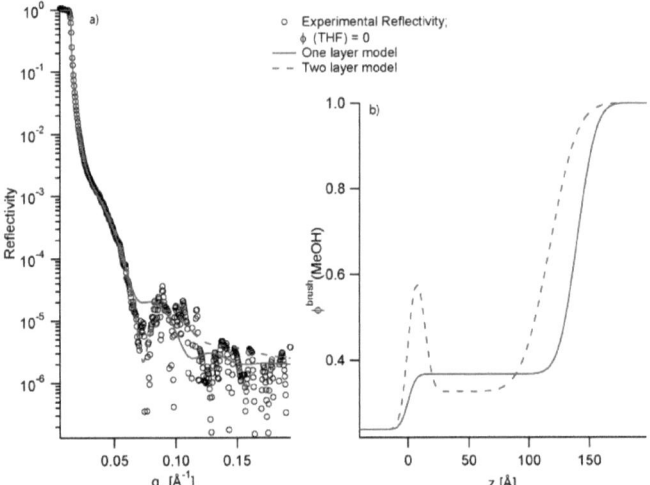

Figure V.6: a) Reflectivity profile of collapsed brush in 100% MeOH of dry origin (left); red lines denote calculated reflectivity profiles for tanh expanded step functions; straight-line: one layer model; dashed line: two layer model; b) corresponding ϕ^{brush} (MeOH) profiles, calculated from SLD profiles according to Eq. (V.3.2)

Stress/structure correlation in grafted from PMMA brushes
-
Neutron reflectivity results

Compared to the reflectivity profile obtained from the collapsed brush of swollen origin Figure V.8 better fits were obtained from a two layer model (dashed line).
The used two layer model suggests preferential MeOH adsorption at the Si/PMMA with a resulting thickness of 15 ± 1Å. Compared to the one layer model, which yielded a brush height of 140 ± 2Å a total brush height of 117 ± 2Å was obtained for the two layer model. Preferential MeOH adsorption at the Si/PMMA interface may be explained by favored interaction energies between polar SiO_x groups with the polar MeOH. This assumption is supported by the observed spreading of MeOH on SiO_x surfaces with contact angles of $\sim 0°$. Such layer is followed by a polymer phase containing only $\phi^{brush}(MeOH) = 0.33 \pm 0.04$, compared to $\phi^{brush}(MeOH + THF) = 0.4 \pm 0.04$ as found for the collapsed brush of swollen origin (Figure V.10). Compared to the collapsed brush of swollen origin the resulting 2nd layer is comparably compact. Urayama et al. roughly estimated the degree of entanglement in stretched polymer brushes[148, 151]. The authors used the ratio of the chains mean-square end-to-end distance $\langle r^2 \rangle^{1/2}$ normal to the substrate in the unstretched thin film case in respect to the stretched brush film for their calculations.

$$\alpha^2 = \left[\frac{1+\left(H/\langle r^2 \rangle^{1/2}\right)}{2\cdot\left(H/\langle r^2 \rangle^{1/2}\right)} \right]$$ (V.3.4)

Eq. (V.3.4) gives a quantitative measure for the percentage change $\Delta\alpha^2 = \alpha^2_{bulk} - \alpha^2_{brush}$ of entanglements compared to an unperturbated Gaussian chain. For the unperturbated Gaussian chain in the bulk polymer, $\alpha^2_{bulk} = 1$ is obtained. For polymer brush films thinner than $\langle r^2 \rangle^{1/2}$, the degree of entanglements is higher compared to the bulk polymer and $\Delta\alpha^2 < 0$ is obtained. In the stretched case the degree of entanglements is higher than in the bulk polymer and $\Delta\alpha^2 > 0$ is obtained.
For the dried PMMA brush with a brush height of 7.1 nm (x-ray reflectivity from Figure V.5) and $\langle r^2 \rangle^{1/2} = 9.8$ nm, $\Delta\alpha^2 = 0.19$ is obtained. The entanglement molecular weight of PMMA in concentrated solutions is ~ 11000 g/mol[152]. Thus, the dry PMMA brush of $M_w = 23500$ g/mol can

be regarded in an entangled state. Immersion in MeOH leads to $\Delta\alpha^2 = -0.08$ for the collapsed brush of dry origin and $\Delta\alpha^2 = -0.15$ for the collapsed brush of swollen origin.

These results would suggest that chain entanglements were formed for both brush collapse routes. However, the degree of entanglement was lower for the collapsed brush of swollen origin. There are two possible reasons for this observation. The collapsed brush of swollen origin could be in a kinetically frozen state and chain entanglements could not be reduced to equilibrium conditions. The other possibility is that the fast collapsed brush from swollen origin is kinetically hindered and unable to retrieve entanglements at equilibrium. This issue was investigated in detail with surface stress investigations presented in chapter V.4.

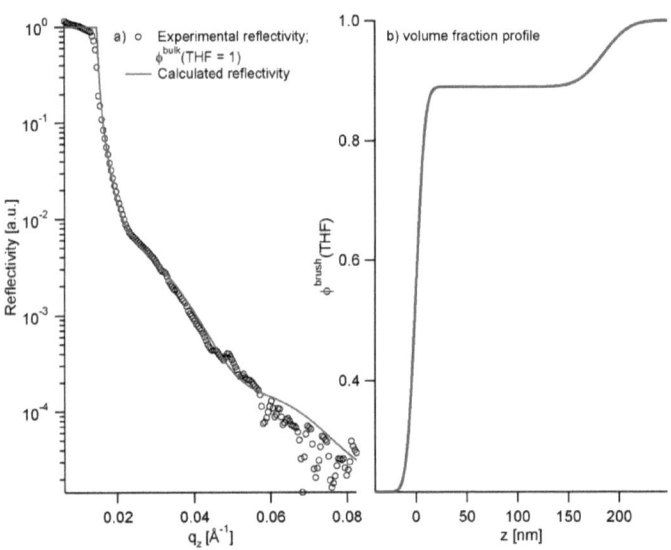

Figure V.7: Reflectivity profile of swollen brush in 100% THF (left); red line denotes calculated reflectivity profiles for tanh expanded step functions; b) corresponding ϕ^{brush} (THF) profile, calculated from SLD profiles according to Eq. (V.3.2)

In successive experiments, the fast swelling process was investigated. The PMMA brush was completely swollen in a solution of 100% THF. After totally swelling to 185 ± 3 Å, a volume fraction of $\phi^{brush}(THF) = 0.89 \pm 0.04$ was found (Figure V.7). Chain entanglements were reduced by

$\Delta\alpha^2 = -0.24$. For comparison, entanglements in the gradual swollen brush were only reduced by $\Delta\alpha^2 = -0.12$ (chapter V.3.4). The higher magnitude of entanglement reduction leads to the consequence that ~ 15% more solvent can be adsorbed into the brush phase. Total $\phi_{tot}^{brush}(THF)$ values for the fast swollen brush are $\phi_{tot}^{brush}(THF) = 0.87$ compared to $\phi_{tot}^{brush}(THF) = 0.73$ for the gradual swollen brush.

V.3.4 Gradual collapse/swelling transition

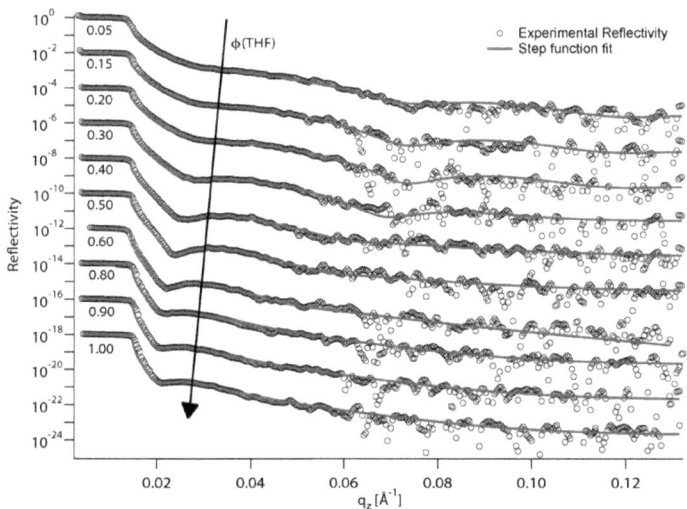

Figure V.8: Reflectivity scans of PMMA brush for various solvent mixtures of MeOH and THF with applied Step function type fit. The bulk volume fraction of THF is indicated at the right hand side of the single graphs. The reflectivity scans were shifted by a factor of 10^{-2} against each other for increasing volume fractions of THF.

In the proceeding experimental series polymer/solvent interaction parameters were deduced from NR results obtained for gradual brush swelling. As mentioned in chapter V.3.1, the PMMA brush was collapsed from the swollen brush, and the bulk's volume fraction of THF in the bulk (ϕ^{bulk}(THF)) was steadily increased, starting from ϕ^{bulk}(THF) = 0.05 (upper graph in Figure V.8).

Neutron reflectivity results

Regarding qualitatively the obtained reflectivity curves it can be directly seen that Fresnel's oscillations move towards smaller q_z for increasing ϕ^{bulk} (THF). This observation is a clear qualitative evidence for a gradual brush thickness increase.

For quantitative data analysis *tanh* expanded step type density models were used to analyze experimental reflectivity profiles. Figure V.8 shows that good fits can be obtained using a one layer for the collapsed and swollen PMMA.

Table V.3.1: Fit results with estimated errors for all reflectivity data presented in Figure V.8.

ϕ^{bulk} (THF)	H (brush) [Å]	SLD^{brush} [10^{-6}Å^{-2}]	SLD^{bulk} [10^{-6}Å^{-2}]	s [Å]
0.05	130 ± 3	2.8 ± 0.1	5.83 ± 0.05	12 ± 2
0.15	130 ± 3	3.0 ± 0.1	5.88 ± 0.05	15 ± 2
0.2	129 ± 3	3.1 ± 0.1	5.91 ± 0.05	16 ± 2
0.3	135 ± 3	3.4 ± 0.1	5.97 ± 0.05	24 ± 3
0.4	143 ± 3	3.5 ± 0.1	6.02 ± 0.05	36 ± 3
0.5	153 ± 3	3.6 ± 0.1	6.08 ± 0.05	40 ± 3
0.6	168 ± 3	3.6 ± 0.1	6.13 ± 0.05	52 ± 5
0.8	170 ± 3	4.7 ± 0.1	6.24 ± 0.05	60 ± 5
0.9	174 ± 3	4.7 ± 0.1	6.30 ± 0.05	68 ± 5
1	175 ± 3	4.6 ± 0.1	6.35 ± 0.05	70 ± 5

From obtained fit values presented in Table V.3.1 corresponding ϕ^{brush} (MeOH + THF) profiles can be calculated (Figure V.9).

Neutron reflectivity results

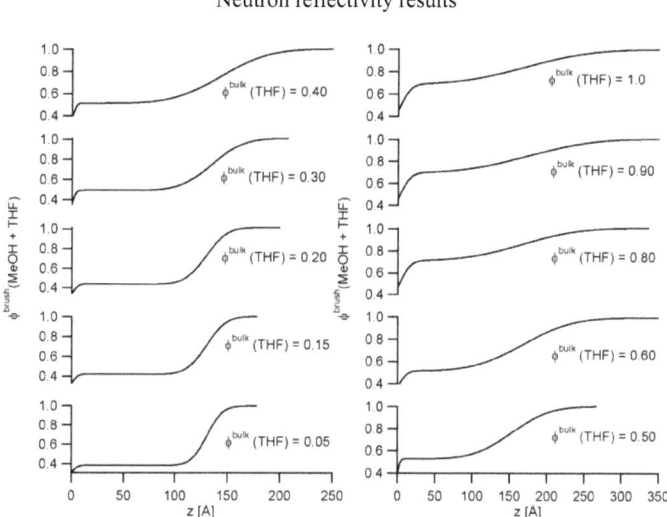

Figure V.9: ϕ^{brush} (MeOH + THF) profiles calculated form fit results presented in Table V.3.1.

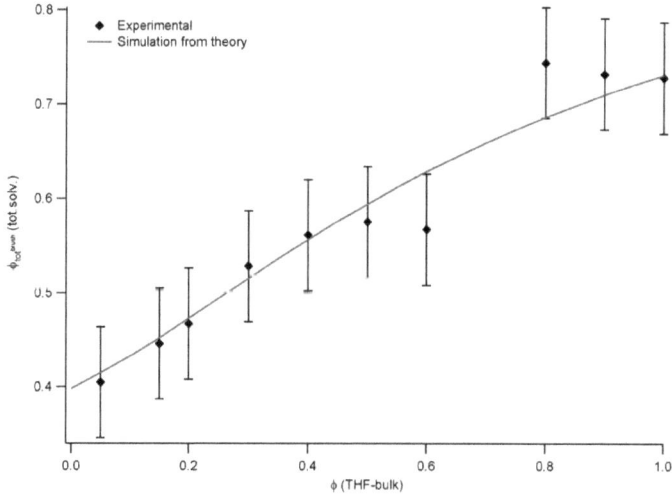

Figure V.10: Plot of deduced total solvent volume fractions in the brush phase vs. the bulk's solvent composition with estimated errors. The straight red line corresponds to theoretical calculations[30] with used Flory Huggins interaction parameters of $\chi_{THF} = -0.3$, $\chi_{MeOH} = 0.8$ and $\chi_{THF/MeOH} = -0.6$.

Stress/structure correlation in grafted from PMMA brushes
-
Neutron reflectivity results

Obtained results for $\phi_{tot}^{brush}(MeOH+THF)$ were fitted with theory from Birshtein and Latskaya[30] (chapter II.3.2.2), which allowed to estimate Flory Huggins interaction parameters χ_{THF}, χ_{MeOH} and $\chi_{THF/MeOH}$. Relevant parameters, which were needed for the simulations are listed in Table V.3.2.

Table V.3.2: Parameters used for χ-parameter estimation

M_w (PMMA)	23500 g/mol
M_w (MMA)	114 g/mol
N	206
Segment length l^{58}	0.66 nm
σ	13[vi]

Best matched simulations were obtained for interaction parameter sets of χ_{THF} = -0.3, χ_{MeOH} = 0.8 and $\chi_{THF/MeOH}$ = -0.6. As seen in Figure V.10 simulations from theory are in good agreement with experimental data from 0 < ϕ^{bulk} (THF) ≤ 0.5. For ϕ^{bulk} (THF) > 0.5 experimental observed saturation of solvent adsorption into the PMMA brush phase is not perfectly reflected by theory. In contrast to brush swelling theory, interaction parameter values of χ_{THF} = 0.36 and χ_{MeOH} = 1.2 would be obtained from Hildebrand-Scott solubility estimations[139].

Interaction parameter estimation using the Hildebrand-Scott theory is only valid for polymer solvent pairs with an enthalpic interaction part $\chi_H \geq 0$. Thus, interaction parameter estimations for attracting polymer/solvent pairs, e.g. PMMA/THF, are not possible within the common solubility approach. In addition, χ_{MeOH} was experimentally found to be by 0.4 lower, than from Hildebrand-Scott estimations.

The physical reason for this difference can be found in the different geometry of the PMMA brush compared to free PMMA chains. For both, the free PMMA and the PMMA brush, polymer/solvent enthalpic contributions to interaction energies are assumed to be equal. Hence, compared to bulk samples, no essential difference in enthalpic χ_H parts is expected. However, compared to the free PMMA, PMMA chains in the dense brush are entropically constrained by the next neighbor chains and by immobilization to the substrate. Such entropic constraints lower the entropy of mixing with the surrounding solvent. Comparing the PMMA brush's entropy of mixing with the free PMMA brush's entropy of mixing it is possible to write

---

[vi] corresponds to a chain density of 0.18 chains/nm²

$$\Delta S_{Mix}^{brush} < \Delta S_{Mix}^{free} \qquad (\text{V}.3.5)$$

However reduced entropies of mixing would increase the found χ-parameters, instead of decreasing them.

Besides the enthalpic interaction energy and the entropy of mixing, there has to be at least one more energetic contribution, which explains the found χ-parameter values. Compared to free PMMA chains, interfacial energies between the PMMA brush and the solvent have to be considered. The surface tensions of THF and MeOH are γ_{lg} (THF) = 27.05 mN/m and γ_{lg} (MeOH) = 22.07 mN/m[153]. From contact angle experiments with H_2O, formamide and toluene on the PMMA brush specimen γ_{sg} (PMMA-brush) = 45.1 ± 0.7 mN/m, which is near the literature value for bulk PMMA of γ_{sg} (PMMA-bulk) = 41.1 mN/m[139], was obtained. Contact angle experiments with MeOH and THF on the same specimen showed a complete wetting of the sample for both liquids with contact angles < 5°, respectively.

With Young's equation

$$\cos\theta = \frac{\gamma_{sv} - \gamma_{sl}}{\gamma_{lv}} \qquad (\text{V}.3.6)$$

γ_{sl} (PMMA-brush/THF) = 13.7 mN/m and γ_{sl} (PMMA-brush/MeOH) = 18.6 mN/m can be obtained. The definition of the free surface energy regarding only interfacial tensions

$$dG^{\sigma} = -Ad\gamma_{sl} \qquad (\text{V}.3.7)$$

shows that G^{σ} becomes negative when the brush is immersed in MeOH and THF. Since assigned γ_{sl} values for THF and MeOH vary only by ~ 5 mN/m, it is reasonable that surface energy effects would lower χ_{THF} and χ_{MeOH} parameters by a similar magnitude compared to the bulk polymer. Based on contact angle results and due to insufficient description of attractive polymer/solvent interactions by Hildebrand and Scott, the experimentally found χ-parameter set is regarded as reasonable.

Stress/structure correlation in grafted from PMMA brushes
-
Neutron reflectivity results

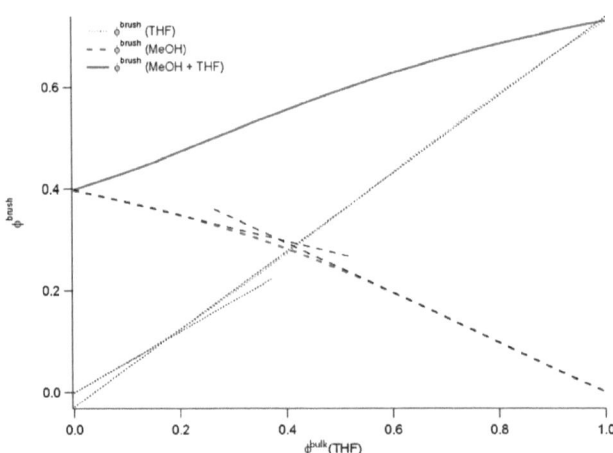

Figure V.11: Simulated ϕ^{brush} (THF) (dotted red line), ϕ^{brush} (MeOH) (dashed red ine) and ϕ^{brush} (MeOH + THF) (straight red line). Black lines allocate for linear regimes for MeOH desorption and THF adsorption.

After discussing the correctness of the experimental assigned χ-parameters, they can be used to simulate single physical ϕ^{brush} (THF) and ϕ^{brush} (MeOH) curves (Figure V.11), which were not directly experimentally accessible. This approach allows the understanding of adsorption and desorption phenomena of the bad solving MeOH and good solving THF. The brush swelling characteristics can be divided into three regimes, according to the turning points of the simulated ϕ^{brush} (MeOH) and ϕ^{brush}(THF) curve.

For ϕ^{bulk} (THF) ≤ 0.2 a constant increase of ϕ^{brush} (MeOH + THF) is observed, while the brush height H = 130Å stays constant. Here ϕ^{brush} (MeOH) decreases with $\frac{d\phi^{brush}(MeOH)}{d\phi^{bulk}(THF)} = -0.24$, while ϕ^{brush} (THF) increases with $\frac{d\phi^{brush}(THF)}{d\phi^{bulk}(THF)} = 0.60$.

Since a constant brush height was observed, it is predicted that THF molecules adsorb preferentially at the brush/solvent interface, by replacing small amounts of MeOH. This prediction is also supported by the increasing brush roughness.

For 0.2 < ϕ^{bulk} (THF) ≤ 0.5 a flattening of the ϕ^{brush} (MeOH + THF) curve is observed, which is accompanied with an increase in H. This behavior is accompanied with an increasing THF adsorption rate of $\frac{d\phi^{brush}(THF)}{d\phi^{bulk}(THF)} = 0.77$, while the MeOH desorption is steadily increasing to $\frac{d\phi^{brush}(MeOH)}{d\phi^{bulk}(THF)} = -0.48$. Thus, MeOH exchange with THF molecules increases, while additional THF adsorption becomes less. It is predicted that THF molecules are able to diffuse into the PMMA brush and replace more and more MeOH molecules. The exchange of MeOH solvent molecules with THF molecules allows the PMMA brush to reduce its conformational entropy by stretching away from the surface.

For ϕ^{bulk} (THF) > 0.5, theory predicts MeOH desorption of $\frac{d\phi^{brush}(MeOH)}{d\phi^{bulk}(THF)} = -0.48$ and THF adsorption of $\frac{d\phi^{brush}(THF)}{d\phi^{bulk}(THF)} = 0.77$. However, saturation observed in experimental data is not sufficiently reflected from theoretical simulations (Figure V.10). Experimental data would predict a saturation of solvent adsorption. Thus, and in contrast to theory, MeOH desorption should equal THF adsorption in order to lead to saturation with constant brush heights.

V.3.5 Summary of neutron reflectivity results

Using neutron reflectivity reasonable polymer solvent interaction parameters could be obtained by fitting experimental results with a theoretical model[30]. The obtained values were found to be lower than parameter sets predicted from Hildebrand-Scott solubility parameter estimations[139]. This result was attributed to complete wetting of the PMMA brush specimen by both solvents. Surface energy effects and attractive polymer/solvent interactions are not included in the Hildebrand-Scott parameter estimations. It is therefore concluded that polymer solvent interaction parameter estimation based on the solubility theory is misleading for polymer brush systems. Using the obtained interaction parameters, the adsorption/desorption behavior of the single solvent components becomes accessible. The adsorption/desorption behavior could be divided into three major regimes. At low bulk concentrations of good solvents, bad solvent molecules are exchanged with good solvent molecules preferably at the brush/liquid interface. For intermediate bulk concentrations of good solvent good solvent molecules were able to diffuse into the whole brush phase with accompanied solvent exchange. For THF fractions reaching unity, saturation of the brush phase with good solvent was observed.

Besides equilibrium aspects, kinetic effects on brush collapse/swelling transitions were observed. It was observed that brush thickness and densities were dependent on the route collapse/swelling transition. This observation was attributed to kinetic entanglement effects. However, it became not clear from NR results, if the collapsed brush of dry (high degree of entanglement) origin is in a kinetically frozen state or if chain entanglements in the collapsed brush of swollen (lower degree of entanglement) origin were unable to be restored. This question was therefore addressed with surface stress investigations, presented in the following chapter.

V.4 Surface stress experiments

At this point chain entanglements effects proposed in the previous chapter are discussed from a mechanical point of view. In this context also the reversibility of the brush mechanics was studied.
In the past extensive experimental and theoretical work was done on elastic stress – strain behavior of entangled polymeric networks[154-156]. In these works the resulting stress on elastic bulk polymer was examined under the influence of an external strain. Later Urayama et al.[148] used the theoretical model derived for bulk networks for the interpretation of mechanics in stretched polymer brushes. The performed approximation was regarded to be reasonable, since the forces causing the brush to stretch away from the surface were interpretated as external strain acting on each polymer brush chain.

Theoretical and experimental works performed on bulk elasticity found decreasing stresses for elongated polymers. In this context elongation or compression is described by the ratio of $L/\langle r^2 \rangle^{1/2}$, where L is the end-to-end distance of the compressed/stretched polymer chain and $\langle r^2 \rangle^{1/2}$ is the end-to-end distance of an unperturbated chain. It was observed that the measured stress reached a minimum at $L \approx 2 \cdot \langle r^2 \rangle^{1/2}$. For $L > 2 \cdot \langle r^2 \rangle^{1/2}$ a steep increase of the measured stress was observed[154, 155]. This effect was observed for natural rubber. However, the trend of stress propagation is used as comparison in order to answer the remaining question of kinetic entanglement effects during collapse/swelling transitions as observed from NR experiments. This was performed in first approximations by replacing L with the brush height H.

Stress/structure correlation in grafted from PMMA brushes
-
Surface stress experiments

V.4.1 Experimental approach

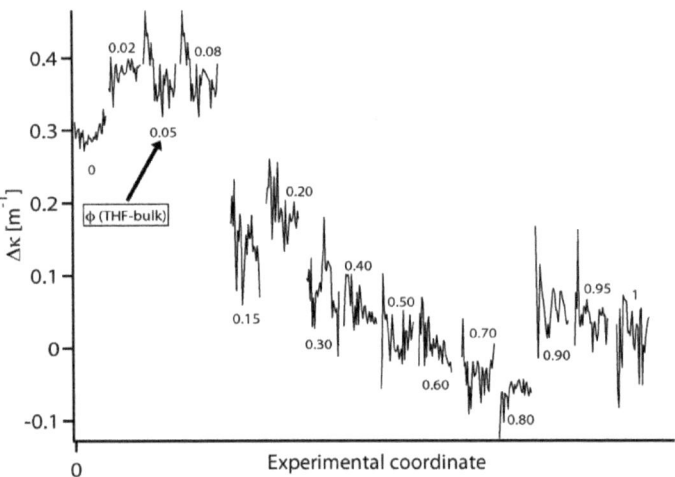

Figure V.12: $\Delta\kappa$ spectra extracted from 3-D curvatures. One spectra yields $\Delta\kappa$ data for each ϕ (THF-bulk) within a time intervall of 30 min.

Surface stress investigations were performed on a similar way than NR experiments. Experiments started from poor (pure MeOH) PMMA solving conditions. Similar to NR studies experimental series were performed with the PMMA brush of dry origin and the PMMA brush of swollen origin. After recording 3-D topographies for 30 min, the bulks volume fraction of THF was increased by total solvent exchange. This procedure was repeated for several ϕ^{bulk} (THF). Thus NR relating curvature and surface stress information was obtained. From obtained MC sensor array 3-D topographies, curvature data was extracted as explained in chapter II.1.7. Averaged curvature data for one topography image was obtained by averaging curvature values from the four PMMA brush coated MC sensors and uncoated reference MC sensors. Differential curvatures $\Delta\kappa$ were obtained by subtraction of reference MC sensor curvatures from PMMA coated MC sensor curvatures (Figure V.12). To obtain average $\Delta\kappa$ values for each ϕ^{bulk} (THF), $\Delta\kappa$ values from each spectrum were averaged and maximum errors were estimated. Relating surface stresses $\Delta\sigma$ were calculated with Stoney's formula Eq. (II.1.3).

V.4.2 Stress propagation for the swelling of the collapsed brush of dry/swollen origin

It can be seen from $\Delta \kappa$ and $\Delta \sigma$ data presented in Figure V.13a that surface stress propagations for the swelling of the collapsed brush of dry origin (black curve) and the brush of swollen origin (red curve) are different.

Surface stress propagation related to the brush collapse/swelling transition with swollen origin an almost constant value of $\Delta \sigma \sim 1$ mN/m for ϕ^{bulk} (THF) ≤ 0.8. For ϕ^{bulk} (THF) > 0.8 the surface stress turns to compressive with $\Delta \sigma \sim -3$ mN/m. Except for high contents of THF, the surface stress did not show the from bulk studies[154, 155] expected surface stress decrease. Regarding bulk theories applicable for the polymer brush geometry[148], surface stress results suggest that after collapsing from the swollen state, the PMMA brush is kinetically hindered to retrieve its predicted entanglements within the given timeframe. It seems that entanglements were only able to be established after some time. In the case of the presented experimental series this was at ϕ^{bulk} (THF) > 0.8.

The situation changes measuring the surface stress propagation for the swelling of the PMMA brush of dry origin. Since the brush was dried under closed air atmosphere over several weeks, entanglements were regarded to be near equilibrium conditions. From the discussion in the previous chapter, the degree of entanglement was even $\sim 20\%$ higher ($\alpha^2 = 1.19$) than in the bulk case. As observed from the surface stress progression the surface stress decays towards a minimum of ~ 0 within the experimental error. Such surface stress decays upon brush stretching are qualitatively reflected by bulk elasticity theory[155, 156].

Stress/structure correlation in grafted from PMMA brushes - Surface stress experiments

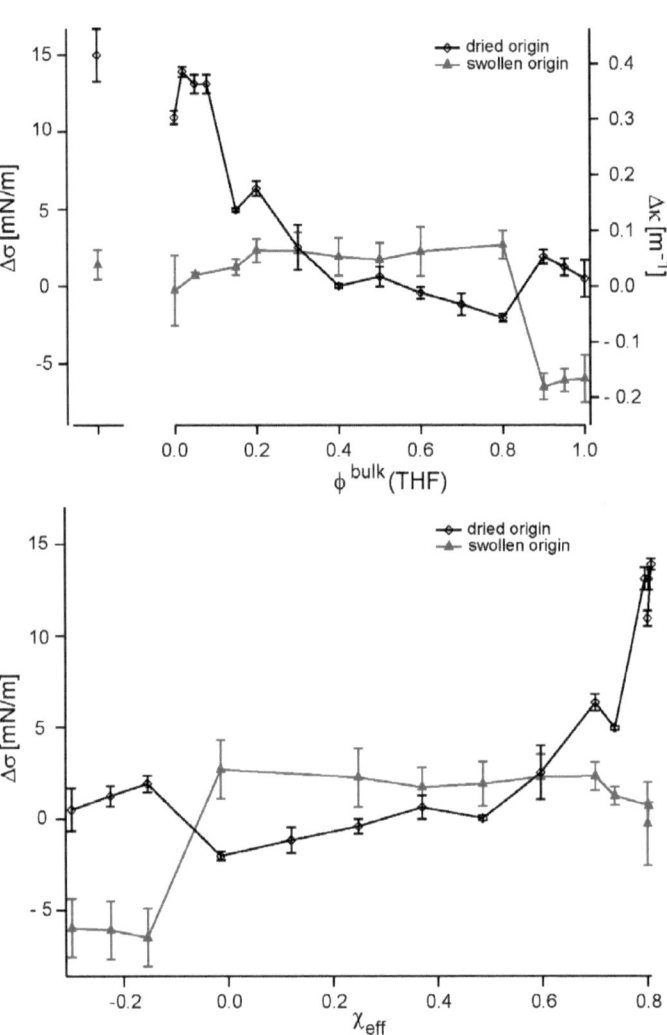

Figure V.13: a) deduced differential curvature $\Delta\kappa$ and differential surface stress $\Delta\sigma$ data in respect to volume of THF in the bulk; b) comparison of $1/\phi^{brush}$ (MeOH + THF) data as obtained from NR data fitting simulations with surface stress results.

Using effective polymer/solvent interaction parameters, χ_{eff}, which were deduced from χ_{MeOH}, χ_{THF} and $\chi_{MeOH/THF}$ (Eq. (II.3.19)), surface stress progression can be related to thermodynamics. Surface stresses reach 0 for χ_{eff} = 0.5 and thus under θ conditions. This result suggests that chain entanglements decrease towards a minimum reaching θ solvent conditions. In the good solvent regime for χ_{eff} < 0.5 only small scattering of data points is observed. Entanglement related mechanics do not change essentially in the good solving regime. Hence, further swelling of the polymer brush network did not change the degree of entanglement.

V.4.3 Summary of surface stress results

Using surface stress experiments it was possible to conclude that collapsed brushes of dry origin were in a highly entangled state, while collapsed brushes of swollen origin were not able to retrieve their degree of entanglement. This conclusion is qualitatively supported by investigations from polymeric bulk networks[154]. In contrast surface stress propagation for the collapsed brush of dried origin showed the from bulk theory[155, 156] predicted surface stress decrease. The surface stress reached a minimum for θ solving conditions. It was therefore concluded that surface stress related entanglements did not change essentially in the good solving regime.

VI Summary and Outlook

In this thesis, studies on 'grafted to' PS/PVME polymer blends and 'grafted from' PMMA polymer brushes were presented. Apart from experimental studies, theoretical effort supported by simulations and experiments was made to obtain a suitable model for the analysis of GISAXS.

For the analysis of performed GISAXS experiments a versatile scattering model was adapted from transmission scattering theory. Supporting on theoretical considerations, comparisons with simulations and experiments showed that quantitative information on structure forms, mean dimensions and domain centre to centre distances could be obtained within theoretical deviations < 20%. Such limitations are usually not crucial for colloidal and polymer systems, since polydispersities within these systems are typically higher than 20%. Future model developments should make quantitative predictions on polydispersities and vertical film structures possible.

Experimental results of PS/PVME blend, which exhibits LCST behavior in none grafted, and homopolymer films grafted to UV sensitive surfaces were shown. A combination of structural investigating techniques, such as surface probe microscopy, µ-x-ray reflectivity and µ-GISAXS together with surface stress investigations was used to characterize the systems.
It was shown that the phase separation behavior of the PS/PVME film could be adjusted by the density of active benzophenone (BP) groups. It was concluded that the grafting point densities is an important parameter for the adjustment of polymer/polymer phase separation processes. Surface stress investigations allowed understanding the molecular chain mechanics of the grafted polymer systems. It was suggested that PVME chains dewet from a mixed BP/PS phase by recovering their conformational chain entropies.
In continuative studies the film thickness dependence of grafted polymer films in respect to the surface coverage of fully active BP layers was studied. It was shown by x-ray reflectivity that higher surface BP coverage leads to general higher film thicknesses. This result was explained by cooperative effects of increased amount of grafted polymeric material and repulsive interactions between the PVME and the BP surface. However, no phase separation was observed. Surface stress experiments showed that phase transitions in the EtOH-BP bulk are qualitatively preserved in

Summary and Outlook

highly densed monolayers. Such phase transitions can be suppressed by grafting polymer onto the BP layer. It is supposed that the grafting of polymer reduces the flexibility of the BP-molecules, which leads to a suppression of the phase transition. In future studies a reproducible way of adjusting the ratio between activated and deactivated BP groups should be developed. One possibility would be the use of mild reducing agents in different concentrations. A possible application for low grafted films would be the grafting of templating block-co-polymers, which could change their morphology depending on the grafting point density. For highly constrained films grafted on fully active BP surfaces of high surface coverage, and when dewetting should be suppressed, one potential application could be their use in ultra thin conducting polymer films.

Grafted from PMMA brushes were synthesized in a simultaneous preparation route on Si-disk specimen used for neutron reflectivity studies and arrays of micromechanical cantilever sensors used for surface stress investigations. Thus equal polymer chain lengths and grafting densities could be assured for both sample types. Neutron reflectivity studies in the presence of methanol (MeOH)/tetrahydrofuran (THF) solvent mixtures revealed an increase in brush thickness and volume fraction of incorporated solvent, for increasing fractions of THF in the bulk mixture. Experimental reflectivity profiles could be modeled with *tanh* expanded step function profiles. Integrated volume fraction data in respect to the THF bulk fraction was compared with simulations from theory and specific polymer/solvent interaction parameters could be assigned. The combination of neutron reflectivity data with surface stress data proved that there are considerable kinetic effects within the brush swelling mechanism. It was concluded that chain entanglements in the dried brush were decreasing during the swelling process. A minimum was reached at θ – conditions. However, and in agreement with previous studies it was found that the surface stress progression, which is dependent on the degree of entanglement, was influenced by the route of the collapse/swelling process. In conclusion the restoration of the entanglement network was found to be kinetically disfavored, when the brush is collapsed from the swollen state.
It was further observed that fast swelling of the PMMA brush from the collapsed state can adsorb ~ 15% more THF than the gradual swollen brush. This observation can be of high interest for brush applications in the field of industrial adsorbents. Further studies on kinetic brush entanglement effects should be conducted in dependence of the polymer brush's polymeric weight.

# VII APPENDIX

VII.1 Optical constants

Table VII.1.1: SLD values for used compounds in film samples for x-ray reflectivity and GISAXS experiments, as obtained from the online scattering length density calculator[149].

Compound	ρ [g/cm^3]	SLD (real) [10^{-6}Å^{-2}]	SLD (imag) [10^{-8}Å^{-2}]
Si	2.33	20.1	46.5
SiO$_2$	2.21	18.9	24.6
Au	19.3	123	121
TiO$_2$	3.89	31.6	15.5
Cl-BP	1	9.08	3.18
EtOH-BP	1	8.92	3.41
PS	1	9.15	1.17
PVME	1	9.38	1.72
PMMA	1	9.18	1.85

Table VII.1.2: Calculated δ, β and α_c values for $\lambda = 1.381$ Å and $\lambda = 1.54$ Å according to Eqn. (II.2.10), (II.2.11).

	$\lambda = 1.381$ Å			$\lambda = 1.54$ Å		
Compound	δ [10^{-6}]	β [10^{-8}]	α_c [°]	δ [10^{-6}]	β [10^{-8}]	α_c [°]
Si	6.10	14.1	0.20	7.59	17.55	0.22
SiO$_2$	5.74	7.47	0.19	7.13	9.29	0.22
Au	37.3	36.7	0.50	46.4	45.67	0.55
TiO$_2$	9.59	4.70	0.25	11.9	5.85	0.28
Cl-BP	2.76	0.97	0.13	3.43	1.20	0.15
EtOH-BP	2.71	1.04	0.13	3.37	1.29	0.15
PS	2.78	0.36	0.14	3.45	0.44	0.15
PVME	2.85	0.52	0.14	3.54	0.65	0.15
PMMA	2.79	0.56	0.14	3.47	0.70	0.15

APPENDIX

Table VII.1.3: SLD and calculated δ and α_c values for used compounds in film samples for neutron reflectivity experiments at λ = 4.26 Å, as obtained from the online scattering length density calculator[149] and calculated from Eq. (II.2.13)

Compound	ρ [g/cm^3]	SLD (real) [10^{-6}Å^{-2}]	SLD (imag) [10^{-8}Å^{-2}]	λ = 4.26 Å δ [10^{-6}]	β [10^{-8}]	α_c [°]
Si	2.33	2.07	0	5.98	0	0.20
SiO$_2$	2.21	3.49	0	10.08	0	0.26
PMMA	1	0.898	0	2.59	0	0.13
d-MeOH	0.89	5.80	0	16.75	0	0.33
d-THF	0.99	6.35	0	18.34	0	0.35

VII.2 Dimensional and mechanic properties of Si MC sensors

Parameter	Value
Width [μm]	90
Length [μm]	500
Thickness, t [μm]	1; 2
Poisson ratio, ν	0.28
Young's modulus, G	130 Gpa

VII.3 Input-files used for IsGISAXS simulations

VII.3.1 Simulation of GISAXS from Au film

```
#########################################
#  GISAXS SIMULATIONS : INPUT PARAMETERS
#########################################

# Base filename
FS_DWBA_5nm_t07Grad
######################### Framework and beam parameters #########################
# Framework  Diffuse, Multilayer, Number of index slices, Polarization
   DWBA      LMA      0      25         ss
# Beam Wavelenght : Lambda(nm), Wl_distribution, Sigma_Wl/Wl, Wl_min(nm), Wl_max(nm), nWl, xWl
          0.138       none     0.3      0.08       0.12     20    3
# Beam Alpha_i  : Alpha_i(deg), Ai_distribution, Sigma_Ai(deg), Ai_min(deg), Ai_max(deg), nAi, xAi
          0.7         none       0.1         0.15        0.25       30    2
# Beam 2Theta_i : 2Theta_i(deg), Ti_distribution, Sigma_Ti(deg), Ti_min(deg), Ti_max(deg), nTi, XTi
          0           none        0.5          -0.5         0.5       10    2
```

APPENDIX

```
# Substrate : n-delta_S, n-beta_S,  Layer thickness(nm), n-delta_L,  n-beta_L,  RMS roughness(nm)
         6.10E-06    1.02e-7      20      3.73E-05   3.67E-06    0.
# Particle : n-delta_I,  n-beta_I,   Depth(nm), n-delta_SH,  n-beta_SH
         4.65E-05    4.57E-06    0     8.E-04    2.e-8
############################### Grid parameters ###################################################
# Ewald mode
  T
# Output angle (deg) :  Two theta min-max, Alphaf min-max,  n(1),  n(2)
              0.01    10    0  0.50      300    1
# Output q(nm-1) :  Qx min-max, Qy min-max, Qz min-max,  n(1), n(2), n(3)
         -1  1    -1  0    -2  0      200 200   1
############################### Particle parameters ###################################################
# Number of different particle types
 1
# Particle type,   Probability
full_sphere         1
# Geometrical parameters : Base angle (deg), Height ratio, Flattening, FS-radii/R
              54.73     1.              1.     0.8 0.8
# Shell thicknesses (nm) : dR,  dH,  dW
              0  0  0
# H_uncoupled, W_uncoupled
     T    T
# Size of particle        : Radius(nm), R_distribution, SigmaR/R,   Rmin(nm),Rmax(nm), nR, xR
              5      none       0.01    0.1    11   100  4
# Height aspect ratio    : Height/R,  H_distribution, SigmaH/H,  Hmin/R,   Hmax/R,  nH, xH,  rho_H
              1      none      0.1    0.1     11    25  2    0
# Width aspect ratio    : Width/R,  W_distribution, SigmaW/W,  Wmin/R,   Wmax/R,  nW, xW,  rho_W
              2      none      0.4    1     300    15 2   0
# Orientation of particle : Zeta(deg), Z_distribution, SigmaZ(deg), Zmin(deg), Zmax(deg),   nZ, xZ
              0     none    20.     0       120   30  2
################################### Lattice parameters ###################################################
# Lattice type
      none
# Interference function :   Peak position D(nm),  w(nm), Statistics, Eta_Voigt, Size-Distance coupling, Cut-off
              20        5    gau     0.5     0       10000000
# Pair correlation function :  Density(nm-2), D1(nm),  sigma(nm)
              0.007      25     100
# Lattice parameters : L(1)(nm), L(2)(nm), Angle(deg,  Xi_fixed
              10    10      90.      F
              Xi(deg), Xi_distribution, SigmaXi(deg), Ximin(deg),  Ximax(deg), nXi, xXi
              0    gate     20    0.      240.    3   -2
              Domain sizes DL(nm), DL_distribution, SigmaDL/DL, DLmin(nm), DLmax(nm), nDL, XDL
              20000  20000    none    0.2 0.2  200 200   10000 10000   10 10 -2 -2
# Imperfect lattice : Rod description,  Rod shape,
              rec_ellip     cau cau
              Correlation lenghts(nm),  Rod orientation(deg)
```

APPENDIX

```
                    3000    1000     0  90
# Paracrystal :  Probability description
                ellip
                Disorder factors w(nm), DL-statistical distribution and rod orientation (deg)
                0.5  0.5  0.5  0.5
                cau  cau  cau  cau
                0  90  0  90
# Pattern :   Regular pattern content,  Number of particles per pattern
                       F           2
                Positions xp/L, Debye-Waller factors B11/L1 B22/L1 B12/L1
                0.   0.   0.   0.   0.
                0.5  0.5  0.   0.   0.
```

VII.3.2 Simulation of GISAXS from TiO2 particles buried in a PMMA film matrix

```
#########################################
#  GISAXS SIMULATIONS : INPUT PARAMETERS
#########################################
# Base filename
FS_DWBA_5nm_t055Grad_Lay40_Dep20
########################### Framework and beam  parameters #########################################
# Framework  Diffuse, Multilayer, Number of index slices, Polarization
   DWBA_LAYER_ISLAND    LMA      0     25      ss
# Beam Wavelenght : Lambda(nm), Wl_distribution, Sigma_Wl/Wl, Wl_min(nm), Wl_max(nm), nWl, xWl
       0.138      none      0.3     0.08     0.12   20   3
# Beam Alpha_i    : Alpha_i(deg), Ai_distribution, Sigma_Ai(deg), Ai_min(deg), Ai_max(deg), nAi, xAi
       0.55       none      0.1     0.15     0.25   30   2
# Beam 2Theta_i   : 2Theta_i(deg), Ti_distribution, Sigma_Ti(deg), Ti_min(deg), Ti_max(deg), nTi, XTi
       0          none      0.5     -0.5     0.5    10   2
# Substrate : n-delta_S, n-beta_S, Layer thickness(nm), n-delta_L, n-beta_L, RMS roughness(nm)
       7.60E-06   1.76e-7   40      3.46E-06  4.42E-09   0.
# Particle : n-delta_I,  n-beta_I,  Depth(nm), n-delta_SH, n-beta_SH
    1.23E-05    6.01E-06      20     8.E-04    2.e-8
############################### Grid parameters #########################################
# Ewald mode
   T
# Output angle (deg) :  Two theta min-max, Alphaf min-max, n(1), n(2)
            0.01    10     0   0.28    300   1
# Output q(nm-1) :  Qx min-max, Qy min-max, Qz min-max, n(1), n(2), n(3)
        -1 1    -1  0    -2  0      200  200  1
############################### Particle parameters #########################################
# Number of different particle types
1
# Particle type,   Probability
full_sphere          1
# Geometrical parameters : Base angle (deg), Height ratio, Flattening, FS-radii/R
            54.73      1.           1.     0.8  0.8
# Shell thicknesses (nm) : dR,  dH,  dW
             0  0  0
# H_uncoupled, W_uncoupled
     T    T
# Size of particle     : Radius(nm), R_distribution, SigmaR/R,   Rmin(nm),Rmax(nm), nR, xR
            5      none    0.4    0.1    11   100  4
# Height aspect ratio  : Height/R, H_distribution, SigmaH/H,  Hmin/R,    Hmax/R,  nH, xH,  rho_H
            1      none    0.1    0.1    11   25  2   0
# Width aspect ratio   : Width/R, W_distribution, SigmaW/W,  Wmin/R, Wmax/R,  nW, xW,  rho_W
            2      none    0.4    1     300    15  2  0
# Orientation of particle : Zeta(deg), Z_distribution, SigmaZ(deg), Zmin(deg), Zmax(deg),         nZ, xZ
            0      none    20.    0     120   30  2
############################### Lattice parameters #########################################
# Lattice type
```

APPENDIX

```
    none
# Interference function :   Peak position D(nm),  w(nm), Statistics, Eta_Voigt, Size-Distance coupling, Cut-off
                30              15        gau         0.5         0             10000000
# Pair correlation function :   Density(nm-2), D1(nm), sigma(nm)
                    0.007          25       100
# Lattice parameters : L(1)(nm), L(2)(nm), Angle(deg, Xi_fixed
               10      10      90.      F
            Xi(deg), Xi_distribution, SigmaXi(deg), Ximin(deg), Ximax(deg), nXi, xXi
              0        gate         20       0.        240.     3    -2
            Domain sizes DL(nm), DL_distribution, SigmaDL/DL, DLmin(nm), DLmax(nm), nDL, XDL
               20000  20000        none       0.2 0.2  200 200   10000 10000  10 10 -2 -2
# Imperfect lattice : Rod description, Rod shape,
              rec_ellip         cau cau
            Correlation lenghts(nm), Rod orientation(deg)
              3000      1000           0 90
# Paracrystal :    Probability description
                ellip
          Disorder factors w(nm), DL-statistical distribution and rod orientation (deg)
              0.5 0.5  0.5 0.5
              cau cau  cau cau
              0 90  0  90
# Pattern :    Regular pattern content,  Number of particles per pattern
                    F                      2
            Positions xp/L, Debye-Waller factors B11/L1 B22/L1 B12/L1
            0.  0.   0.   0.  0.
                0.5   0.5   0.  0.  0.
```

VII.3.3 Automatic IGOR Pro script for MC bending data analysis

```
#pragma rtGlobals=1      // Use modern global access method.
Macro MCS_40_MeOH_THF_0_100()
variable filenumber, cur_new_point, newpoint,linenumber, cantinumber, b

killwaves/A/Z
cantinumber = 1
newpoint = 0
string sffix = "MeOH_THF_0_100"

do
make /D/N=30 /O $"Curv_C"+num2istr(cantinumber)
make /D/N=30 /O $"TimeFactor_"+Sffix
make /D/N=30 /O $"Time_"+Sffix

cantinumber = cantinumber + 1
while (cantinumber < 9)

filenumber=1

cantinumber = 1

do
GetFileFolderInfo /Z"D:Measurements:Interferometry:1um_MCS_40:"+sffix+":MCS_"+sffix+"_00"+num2istr(filenumber)+".FAE"

$"Time_"+Sffix [newpoint] = (V_modificationDate)

        do
                     linenumber = 1

                    SetFileFolderInfo
/Z"D:Measurements:Interferometry:1um_MCS_40:"+sffix+":topolines:MCS_"+sffix+"_00"+num2istr(filenumber)+"-
CANTI"+num2istr(cantinumber)+"_"+num2istr(linenumber)+".txt"

                    print V_Flag
                    if( V_Flag != 0 )      // file exists

                $"Curv_C"+num2istr(cantinumber) [newpoint] = 0
```

APPENDIX

```
                        else
                            do
                                loadwave/q/g/A
"D:Measurements:Interferometry:1um_MCS_40:"+sffix+":topolines:MCS_"+sffix+"_00"+num2istr(filenumber)+"-
CANTI"+num2istr(cantinumber)+"_"+num2istr(linenumber)+".txt"

                                duplicate /o wave0, $"mean_y_C"+num2istr(cantinumber)+"_F"+num2istr(filenumber)
                                //um column1 zu nennen

                                duplicate /o wave1, $"w_"+num2istr(linenumber)
                                //w1 linienscan 1 für einen canti
                                KillWaves wave0, wave1
                                //(usw.. je nachdem wieviele columns du hast
                                linenumber = linenumber +1

                            while (linenumber < 7)

                            make /D/N =273 /O $"mean_z_C"+num2istr(cantinumber)+"_F"+num2istr(filenumber)
                            $"mean_z_C"+num2istr(cantinumber)+"_F"+num2istr(filenumber) =
(w_1+w_2+w_3+w_4+w_5+w_6)/6
                            CurveFit/NTHR=0/TBOX=0 poly 3, $"mean_z_C"+num2istr(cantinumber)+"_F"+num2istr(filenumber)
[45,184] /X= $"mean_y_C"+num2istr(cantinumber)+"_F"+num2istr(filenumber) /D
                            cur_new_point = $"W_coef" [2]
                            $"Curv_C"+num2istr(cantinumber) [newpoint] = cur_new_point
                            $"TimeFactor_"+Sffix [newpoint] = filenumber

                            Killwaves $"mean_z_C"+num2istr(cantinumber)+"_F"+num2istr(filenumber),
$"mean_y_C"+num2istr(cantinumber)+"_F"+num2istr(filenumber)
                        endif

                        cantinumber = cantinumber +1

            while (cantinumber<9)

            cantinumber = 1
            newpoint = newpoint +1

filenumber = filenumber +1

while (filenumber<10)

// ----> Processing files 10 -100

cantinumber = 1
do
GetFileFolderInfo /Z"D:Measurements:Interferometry:1um MCS 40:"+sffix+":MCS_"+sffix+"_0"+num2istr(filenumber)+".FAE"

$"Time_"+Sffix [newpoint] = (V_modificationDate)
                        do

                            linenumber = 1
                            SetFileFolderInfo
/Z"D:Measurements:Interferometry:1um_MCS_40:"+sffix+":topolines:MCS_"+sffix+"_0"+num2istr(filenumber)+"-
CANTI"+num2istr(cantinumber)+"_"+num2istr(linenumber)+".txt"
                            print V_Flag
                            if( V_Flag != 0 )      // file exists
                            $"Curv_C"+num2istr(cantinumber) [newpoint] = 0

                        else
                            do
                                loadwave/q/g/A
"D:Measurements:Interferometry:1um_MCS_40:"+sffix+":topolines:MCS_"+sffix+"_0"+num2istr(filenumber)+"-
CANTI"+num2istr(cantinumber)+"_"+num2istr(linenumber)+".txt"

                                duplicate /o wave0, $"mean_y_C"+num2istr(cantinumber)+"_F"+num2istr(filenumber)
```

APPENDIX

```
                                //um column1 zu nennen

                                duplicate /o wave1, $"w_"+num2istr(linenumber)
                                //w1 linienscan 1 für einen canti
                                KillWaves wave0, wave1
                                //(usw.. je nachdem wieviele columns du hast
                                linenumber = linenumber +1

                                while (linenumber < 7)

                                make /D/N =273 /O $"mean_z_C"+num2istr(cantinumber)+"_F"+num2istr(filenumber)
                                $"mean_z_C"+num2istr(cantinumber)+"_F"+num2istr(filenumber) =
(w_1+w_2+w_3+w_4+w_5+w_6)/6
                                CurveFit/NTHR=0/TBOX=0 poly 3, $"mean_z_C"+num2istr(cantinumber)+"_F"+num2istr(filenumber)
[45,184] /X= $"mean_y_C"+num2istr(cantinumber)+"_F"+num2istr(filenumber) /D
                                cur_new_point = $"W_coef" [2]
                                $"Curv_C"+num2istr(cantinumber) [newpoint] = cur_new_point
                                $"TimeFactor_"+Sffix [newpoint] = filenumber

                                Killwaves $"mean_z_C"+num2istr(cantinumber)+"_F"+num2istr(filenumber),
$"mean_y_C"+num2istr(cantinumber)+"_F"+num2istr(filenumber)
                                endif

                                cantinumber = cantinumber +1

                                while (cantinumber<9)

            cantinumber = 1
            newpoint = newpoint +1

filenumber = filenumber +1

while (filenumber<31)


Curv_C1 = -Curv_C1/1000
Curv_C2 = -Curv_C2/1000
Curv_C3 = -Curv_C3/1000
Curv_C4 = -Curv_C4/1000
Curv_C5 = -Curv_C5/1000
Curv_C6 = -Curv_C6/1000
Curv_C7 = -Curv_C7/1000
Curv_C8 = -Curv_C8/1000

cantinumber = 1

Edit $"TimeFactor_"+Sffix;DelayUpdate

do

duplicate/O $"Curv_C"+num2istr(cantinumber), $"Curv_C_"+num2istr(cantinumber)+"_"+Sffix

AppendToTable $"Curv_C_"+num2istr(cantinumber)+"_"+Sffix

cantinumber = cantinumber +1

while (cantinumber < 9)

duplicate/O $"Curv_C_1_"+Sffix, $"C_unc_"+Sffix
duplicate/O $"Curv_C_1_"+Sffix, $"C_coat_"+Sffix
duplicate/O $"Curv_C_1_"+Sffix, $"Diff_Curv_"+Sffix


$"C_unc_"+sffix = (Curv_C1 + Curv_C2 + Curv_C3 + Curv_C4)/4
$"C_coat_"+Sffix = (Curv_C5 + Curv_C6 + Curv_C7 + Curv_C8)/4

$"Diff_Curv_"+Sffix = ((Curv_C5 + Curv_C6 + Curv_C7 + Curv_C8)/4) - ((Curv_C1 + Curv_C2 + Curv_C3 + Curv_C4)/4)

newpoint = 0
```

APPENDIX

```
do

$"Time_"+Sffix [newpoint] = ($"Time_"+Sffix [newpoint] - Start_Time [0])/60
newpoint = newpoint +1
while (newpoint <30)

AppendToTable $"C_unc_"+sffix, $"C_coat_"+Sffix, $"Diff_Curv_"+Sffix
AppendToTable $"Time_"+Sffix
Killwaves/A/Z

Endmacro

make /D/N =185 /O wave0, wave1
                                    wave0 = 0
                                    wave1 = 0

                                    duplicate /O wave0, $"mean_y_C"+num2istr(cantinumber)+"_F"+num2istr(filenumber)
                                    duplicate /O wave1, $"w_"+num2istr(linenumber)
newpoint = 0
b = Time_MeOH_THF_100_0 [0]
do

$"Time_"+Sffix [newpoint] = $"Time_"+Sffix [newpoint] - b
newpoint = newpoint +1
while (newpoint <30)
```

References

1. Small, D. J.; Courtney, P. J., Fundamentals of industrial adhesives. *Advanced Materials & Processes* **2005,** 163, (5), 44-47.

2. Pizzi, A.; Mittal, K. L., *Handbook of adhesive technology*. Dekker: New York (u.a.), 1994; p XI,680 S.

3. Raviv, U.; Giasson, S.; Kampf, N.; Gohy, J. F.; Jerome, R.; Klein, J., Lubrication by charged polymers. *Nature* **2003,** 425, (6954), 163-165.

4. Ma, M.; Hill, R. M., Superhydrophobic surfaces. *Current Opinion in Colloid & Interface Science* **2006,** 11, (4), 193-202.

5. Blossey, R., Self-cleaning surfaces [mdash] virtual realities. *Nat Mater* **2003,** 2, (5), 301-306.

6. Ho Lee, C.; Soon An, D.; Cheol Lee, S.; Jin Park, H.; Sun Lee, D., A coating for use as an antimicrobial and antioxidative packaging material incorporating nisin and [alpha]-tocopherol. *Journal of Food Engineering* **2004,** 62, (4), 323-329.

7. Ertl, G., Reactions at surfaces: From atoms to complexity (Nobel lecture). *Angewandte Chemie-International Edition* **2008,** 47, (19), 3524-3535.

8. Matheson, R. R., Paint and coatings technology: Current industrial trends. *Polymer Reviews* **2006,** 46, (4), 341-346.

9. Smith, K. M.; Fowler, G. D.; Pullket, S.; Graham, N. J. D., Sewage sludge-based adsorbents: A review of their production, properties and use in water treatment applications. *Water Research* **2009,** 43, (10), 2569-2594.

10. Choy, K. L., Chemical vapour deposition of coatings. *Progress in Materials Science* **2003,** 48, (2), 57-170.

11. Harper, J. M. E.; Cuomo, J. J.; Kaufman, H. R., Technology and Applications of Broad-Beam Ion Sources Used in Sputtering .2. Applications. *Journal of Vacuum Science & Technology* **1982,** 21, (3), 737-756.

12. Kaufman, H. R.; Cuomo, J. J.; Harper, J. M. E., Technology and Applications of Broad-Beam Ion Sources Used in Sputtering .1. Ion-Source Technology. *Journal of Vacuum Science & Technology* **1982,** 21, (3), 725-736.

References

13. Greener, Y.; Middlema.S, Blade-Coating of a Viscoelastic Fluid. *Polymer Engineering and Science* **1974**, 14, (11), 791-796.

14. Norrman, K.; Ghanbari-Siahkali, A.; Larsen, N. B., Studies of spin-coated polymer films. *Annual Reports on the Progress of Chemistry, Vol 101, Section C, Physical Chemistry* **2005**, 101, 174-201.

15. Hamley, I. W., *Developments in block copolymer science and technology*. Wiley: Chichester [u.a.], 2004; p IX, 367 S.

16. Kim, D. H.; Jia, X. Q.; Lin, Z. Q.; Guarini, K. W.; Russell, T. P., Growth of silicon oxide in thin film block copolymer scaffolds. *Advanced Materials* **2004**, 16, (8), 702-+.

17. Kim, D. H.; Kim, S. H.; Lavery, K.; Russell, T. P., Inorganic nanodots from thin films of block copolymers. *Nano Letters* **2004**, 4, (10), 1841-1844.

18. Prucker, O.; Naumann, C. A.; Ruhe, J.; Knoll, W.; Frank, C. W., Photochemical attachment of polymer films to solid surfaces via monolayers of benzophenone derivatives. *Journal of the American Chemical Society* **1999**, 121, (38), 8766-8770.

19. Bank, M.; Leffingw.J; Thies, C., Influence of Solvent Upon Compatibility of Polystyrene and Poly(Vinyl Methyl Ether). *Macromolecules* **1971**, 4, (1), 43-46.

20. Nishi, T.; Wang, T. T.; Kwei, T. K., Thermally Induced Phase Separation Behavior of Compatible Polymer Mixtures. *Macromolecules* **1975**, 8, (2), 227-234.

21. Nishi, T.; Kwei, T. K., Cloud Point Curves for Polyvinyl Methyl-Ether) and Monodisperse Polystyrene Mixtures. *Polymer* **1975**, 16, (4), 285-290.

22. Roth, S. V.; Burghammer, M.; Riekel, C.; Muller-Buschbaum, P.; Diethert, A.; Panagiotou, P.; Walter, H., Self-assembled gradient nanoparticle-polymer multilayers investigated by an advanced characterization method: microbeam grazing incidence x-ray scattering. *Applied Physics Letters* **2003**, 82, (12), 1935-1937.

23. Wolkenhauer, M.; Bumbu, G. G.; Cheng, Y.; Roth, S. V.; Gutmann, J. S., Investigation of micromechanical cantilever sensors with microfocus grazing incidence small-angle X-ray scattering. *Applied Physics Letters* **2006**, 89, (5).

24. Beaucage, G., Approximations leading to a unified exponential power-law approach to small-angle scattering. *Journal of Applied Crystallography* **1995**, 28, 717-728.

25. Beaucage, G., Small-angle scattering from polymeric mass fractals of arbitrary mass-fractal dimension. *Journal of Applied Crystallography* **1996**, 29, 134-146.

References

26. Alexander, S., Adsorption of Chain Molecules with a Polar Head a-Scaling Description. *Journal De Physique* **1977,** 38, (8), 983-987.

27. de Gennes, P. G., Conformations of Polymers Attached to an Interface. *Macromolecules* **1980,** 13, (5), 1069-1075.

28. Matyjaszewski, K.; Miller, P. J.; Shukla, N.; Immaraporn, B.; Gelman, A.; Luokala, B. B.; Siclovan, T. M.; Kickelbick, G.; Vallant, T.; Hoffmann, H.; Pakula, T., Polymers at interfaces: Using atom transfer radical polymerization in the controlled growth of homopolymers and block copolymers from silicon surfaces in the absence of untethered sacrificial initiator. *Macromolecules* **1999,** 32, (26), 8716-8724.

29. Matyjaszewski, K.; Xia, J. H., Atom transfer radical polymerization. *Chemical Reviews* **2001,** 101, (9), 2921-2990.

30. Birshtein, T. M.; Lyatskaya, Y. V., Theory of the Collapse-Stretching Transition of a Polymer Brush in a Mixed-Solvent. *Macromolecules* **1994,** 27, (5), 1256-1266.

31. Lang, H. P.; Berger, R.; Andreoli, C.; Brugger, J.; Despont, M.; Vettiger, P.; Gerber, C.; Gimzewski, J. K.; Ramseyer, J. P.; Meyer, E.; Guntherodt, H. J., Sequential position readout from arrays of micromechanical cantilever sensors. *Applied Physics Letters* **1998,** 72, (3), 383-385.

32. Stoney, G. G., The tension of metallic films deposited by electrolysis. *Proceedings of the Royal Society of London Series a-Containing Papers of a Mathematical and Physical Character* **1909,** 82, (553), 172-175.

33. Klein, C. A., How accurate are Stoney's equation and recent modifications. *Journal of Applied Physics* **2000,** 88, (9), 5487-5489.

34. Townsend, P. H.; Brunner, T. A., Elastic Relationships in Layered Composite Media with Approximation for the Case of Thin-Films on a Thick Substrate. *Journal of Applied Physics* **1987,** 62, (11), 4438-4444.

35. Schafer, J. D.; Nafe, H.; Aldinger, F., Macro- and microstress analysis in sol-gel derived $Pb(Zr_xTi_{1-x})O_3$ thin films. *Journal of Applied Physics* **1999,** 85, (12), 8023-8031.

36. Rats, D.; Bimbault, L.; Vandenbulcke, L.; Herbin, R.; Badawi, K. F., Crystalline Quality and Residual-Stresses in Diamond Layers by Raman and X-Ray-Diffraction Analyses. *Journal of Applied Physics* **1995,** 78, (8), 4994-5001.

References

37. Barnes, J. R.; Stephenson, R. J.; Welland, M. E.; Gerber, C.; Gimzewski, J. K., Photothermal Spectroscopy with Femtojoule Sensitivity Using a Micromechanical Device. *Nature* **1994**, 372, (6501), 79-81.

38. Gimzewski, J. K.; Gerber, C.; Meyer, E.; Schlittler, R. R., Observation of a Chemical-Reaction Using a Micromechanical Sensor. *Chemical Physics Letters* **1994**, 217, (5-6), 589-594.

39. Berger, R.; Gerber, C.; Gimzewski, J. K.; Meyer, E.; Guntherodt, H. J., Thermal analysis using a micromechanical calorimeter. *Applied Physics Letters* **1996**, 69, (1), 40-42.

40. Butt, H.-J., A sensitive method to measure changes in the surface stress of solids. *Journal of Colloid and Interface Science* **1996**, 180, (1), 251-260.

41. Berger, R.; Delamarche, E.; Lang, H. P.; Gerber, C.; Gimzewski, J. K.; Meyer, E.; Guntherodt, H. J., Surface stress in the self-assembly of alkanethiols on gold. *Science* **1997**, 276, (5321), 2021-2024.

42. Bergese, P.; Oliviero, G.; Alessandri, I.; Depero, L. E., Thermodynamics of mechanical transduction of surface confined receptor/ligand reactions. *Journal of Colloid and Interface Science* **2007**, 316, 1017-1022.

43. Bhushan, B., *Springer Handbook of Nanotechnology*. 2nd. ed.; 2007.

44. Lang, H. P.; Gerber, C., Microcantilever sensors. *Stm and Afm Studies on (Bio)Molecular Systems: Unravelling the Nanoworld* **2008**, 285, 1-27.

45. Bietsch, A.; Hegner, M.; Lang, H. P.; Gerber, C., Inkjet deposition of alkanethiolate monolayers and DNA oligonucleotides on gold: Evaluation of spot uniformity by wet etching. *Langmuir* **2004**, 20, (12), 5119-5122.

46. Bietsch, A.; Zhang, J. Y.; Hegner, M.; Lang, H. P.; Gerber, C., Rapid functionalization of cantilever array sensors by inkjet printing. *Nanotechnology* **2004**, 15, (8), 873-880.

47. Nett, S. K.; Kircher, G.; Gutmann, J. S., PMMA Brushes Prepared in an Ionic Liquid. *Macromolecular Chemistry and Physics* **2009**, 210, (11), 971-976.

48. Deegan, R. D.; Bakajin, O.; Dupont, T. F.; Huber, G.; Nagel, S. R.; Witten, T. A., Capillary flow as the cause of ring stains from dried liquid drops. *Nature* **1997**, 389, (6653), 827-829.

49. Deegan, R. D., Pattern formation in drying drops. *Physical Review E* **2000**, 61, (1), 475-485.

50. Deegan, R. D.; Bakajin, O.; Dupont, T. F.; Huber, G.; Nagel, S. R.; Witten, T. A., Contact line deposits in an evaporating drop. *Physical Review E* **2000**, 62, (1), 756-765.

References

51. Lavrik, N. V.; Sepaniak, M. J.; Datskos, P. G., Cantilever transducers as a platform for chemical and biological sensors. *Review of Scientific Instruments* **2004,** 75, (7), 2229-2253.

52. Singamaneni, S.; LeMieux, M. C.; Lang, H. P.; Gerber, C.; Lam, Y.; Zauscher, S.; Datskos, P. G.; Lavrik, N. V.; Jiang, H.; Naik, R. R.; Bunning, T. J.; Tsukruk, V. V., Bimaterial microcantilevers as a hybrid sensing platform. *Advanced Materials* **2008,** 20, (4), 653-680.

53. Cesaro-Tadic, S.; Dernick, G.; Juncker, D.; Buurman, G.; Kropshofer, H.; Michel, B.; Fattinger, C.; Delamarche, E., High-sensitivity miniaturized immunoassays for tumor necrosis factor a using microfluidic systems. *Lab on a Chip* **2004,** 4, (6), 563-569.

54. Baller, M. K.; Lang, H. P.; Fritz, J.; Gerber, C.; Gimzewski, J. K.; Drechsler, U.; Rothuizen, H.; Despont, M.; Vettiger, P.; Battiston, F. M.; Ramseyer, J. P.; Fornaro, P.; Meyer, E.; Guntherodt, H. J., A cantilever array-based artificial nose. *Ultramicroscopy* **2000,** 82, (1-4), 1-9.

55. McKendry, R.; Zhang, J. Y.; Arntz, Y.; Strunz, T.; Hegner, M.; Lang, H. P.; Baller, M. K.; Certa, U.; Meyer, E.; Guntherodt, H. J.; Gerber, C., Multiple label-free biodetection and quantitative DNA-binding assays on a nanomechanical cantilever array. *Proceedings of the National Academy of Sciences of the United States of America* **2002,** 99, (15), 9783-9788.

56. Fritz, J.; Baller, M. K.; Lang, H. P.; Strunz, T.; Meyer, E.; Guntherodt, H. J.; Delamarche, E.; Gerber, C.; Gimzewski, J. K., Stress at the solid-liquid interface of self-assembled monolayers on gold investigated with a nanomechanical sensor. *Langmuir* **2000,** 16, (25), 9694-9696.

57. Bumbu, G. G.; Kircher, G.; Wolkenhauer, M.; Berger, R.; Gutmann, J. S., Synthesis and characterization of polymer brushes on micromechanical cantilevers. *Macromolecular Chemistry and Physics* **2004,** 205, (13), 1713-1720.

58. Bumbu, G. G.; Wolkenhauer, M.; Kircher, G.; Gutmann, J. S.; Berger, R., Micromechanical cantilever technique: A tool for investigating the swelling of polymer brushes. *Langmuir* **2007,** 23, (4), 2203-2207.

59. Bhushan, B., Nanomechanical Cantilever Array Sensors. In *Springer Handbook of Nanotechnology*, Springer: 2007; pp 443-459.

60. Helm, M.; Servant, J. J.; Saurenbach, F.; Berger, R., Read-out of micromechanical cantilever sensors by phase shifting interferometry. *Applied Physics Letters* **2005,** 87, (6).

61. Korsunsky, A. M.; Cherian, S.; Raiteri, R.; Berger, R., On the micromechanics of micro-cantilever sensors: Property analysis and eigenstrain modeling. *Sensors and Actuators a-Physical* **2007,** 139, (1-2), 70-77.

References

62. Russell, T. P., X-ray and neutron reflectivity for the investigation of polymers. *Materials Science Reports* **1990**, 5, (4), 171-271.

63. Tolan, M., *X-ray scattering from soft-matter thin filmsmaterials science and basic research.* Springer: Berlin [u.a.], 1999.

64. Stamm, M.; Huttenbach, S.; Reiter, G.; Springer, T., Initial-Stages of Polymer Interdiffusion Studied by Neutron Reflectometry. *Europhysics Letters* **1991**, 14, (5), 451-456.

65. Stamm, M.; Majkrzak, C. F., Investigation of Polymeric Thin-Films by Total Reflection of Neutrons. *Abstracts of Papers of the American Chemical Society* **1987**, 194, 30-POLY.

66. Stearns, D. G., The Scattering of X-Rays from Nonideal Multilayer Structures. *Journal of Applied Physics* **1989**, 65, (2), 491-506.

67. Schubert, D. W. Dissertation, Universität Mainz, 1996.

68. Milner, S. T.; Witten, T. A., Bending Moduli of Polymeric Surfactant Interfaces. *Journal De Physique* **1988**, 49, (11), 1951-1962.

69. Zhulina, E. B.; Borisov, O. V.; Priamitsyn, V. A., Theory of Steric Stabilization of Colloid Dispersions by Grafted Polymers. *Journal of Colloid and Interface Science* **1990**, 137, (2), 495-511.

70. Mandelbrot, B. B., *The Fractal Geometry of Nature.* Freeman: New York, 1982.

71. Andelman, D.; Joanny, J. F.; Robbins, M. O., Complete Wetting on Rough Surfaces - Statics. *Europhysics Letters* **1988**, 7, (8), 731-736.

72. Buff, F. P.; Lovett, R. A.; Stilling.Fh, Interfacial Density Profile for Fluids in Critical Region. *Physical Review Letters* **1965**, 15, (15), 621-&.

73. Robbins, M. O.; Andelman, D.; Joanny, J. F., Thin Liquid-Films on Rough or Heterogeneous Solids. *Physical Review A* **1991**, 43, (8), 4344-4354.

74. Fredrickson, G. H.; Ajdari, A.; Leibler, L.; Carton, J. P., Surface-Modes and Deformation Energy of a Molten Polymer Brush. *Macromolecules* **1992**, 25, (11), 2882-2889.

75. Sinha, S. K.; Sirota, E. B.; Garoff, S.; Stanley, H. B., X-Ray and Neutron-Scattering from Rough Surfaces. *Physical Review B* **1988**, 38, (4), 2297-2311.

76. Spiller, E.; Stearns, D.; Krumrey, M., Multilayer X-Ray Mirrors - Interfacial Roughness, Scattering, and Image Quality. *Journal of Applied Physics* **1993**, 74, (1), 107-118.

77. Savage, D. E.; Kleiner, J.; Schimke, N.; Phang, Y. H.; Jankowski, T.; Jacobs, J.; Kariotis, R.; Lagally, M. G., Determination of Roughness Correlations in Multilayer Films for X-Ray Mirrors. *Journal of Applied Physics* **1991**, 69, (3), 1411-1424.

References

78. Holy, V.; Baumbach, T., Nonspecular X-Ray Reflection from Rough Multilayers. *Physical Review B* **1994**, 49, (15), 10668-10676.

79. Holy, V.; Kubena, J.; Ohlidal, I.; Lischka, K.; Plotz, W., X-Ray Reflection from Rough Layered Systems. *Physical Review B* **1993**, 47, (23), 15896-15903.

80. Kortright, J. B., Nonspecular X-Ray-Scattering from Multilayer Structures. *Journal of Applied Physics* **1991**, 70, (7), 3620-3625.

81. Müller-Buschbaum, P.; Bauer, E.; Maurer, E.; Roth, S. V.; Gehrke, R.; Burghammer, M.; Riekel, C., Large-scale and local-scale structures in polymerblend films: a grazing-incidence ultra-small-angle X-ray scattering and sub-microbeam grazing-incidence small-angle X-ray scattering investigation. *Journal of Applied Crystallography* **2007**, 40, S341-S345.

82. Müller-Buschbaum, P.; Casagrande, M.; Gutmann, J.; Kuhlmann, T.; Stamm, M.; von Krosigk, G.; Lode, U.; Cunis, S.; Gehrke, R., Determination of micrometer length scales with an X-ray reflection ultra small-angle scattering set-up. *Europhysics Letters* **1998**, 42, (5), 517-522.

83. Lengeler, B.; Schroer, C. G.; Kuhlmann, M.; Benner, B.; Guenzler, T. F.; Kurapova, O.; Zontone, F.; Snigirev, A.; Snigireva, I., Refractive x-ray lenses. *Journal of Physics D-Applied Physics* **2005**, 38, (10A), A218-A222.

84. Roth, S. V.; Dohrmann, R.; Dommach, M.; Kuhlmann, M.; Kroger, I.; Gehrke, R.; Walter, H.; Schroer, C.; Lengeler, B.; Muller-Buschbaum, P., Small-angle options of the upgraded ultrasmall-angle x-ray scattering beamline BW4 at HASYLAB. *Review of Scientific Instruments* **2006**, 77, (8).

85. Yoneda, Y., Anomalous Surface Reflection of X Rays. *Physical Review* **1963**, 131, (5), 2010-&.

86. Vineyard, G. H., Grazing-Incidence Diffraction and the Distorted-Wave Approximation for the Study of Surfaces. *Physical Review B* **1982**, 26, (8), 4146-4159.

87. Salditt, T.; Metzger, T. H.; Peisl, J.; Reinker, B.; Moske, M.; Samwer, K., Determination of the Height-Height Correlation-Function of Rough Surfaces from Diffuse-X-Ray Scattering. *Europhysics Letters* **1995**, 32, (4), 331-336.

88. Salditt, T.; Metzger, T. H.; Brandt, C.; Klemradt, U.; Peisl, J., Determination of the Static Scaling Exponent of Self-Affine Interfaces by Nonspecular X-Ray-Scattering. *Physical Review B* **1995**, 51, (9), 5617-5627.

References

89. Rauscher, M.; Salditt, T.; Spohn, H., Small-angle x-ray scattering under grazing incidence: The cross section in the distorted-wave Born approximation. *Physical Review B* **1995**, 52, (23), 16855-16863.

90. Rauscher, M.; Paniago, R.; Metzger, H.; Kovats, Z.; Domke, J.; Peisl, J.; Pfannes, H. D.; Schulze, J.; Eisele, I., Grazing incidence small angle x-ray scattering from free-standing nanostructures. *Journal of Applied Physics* **1999**, 86, (12), 6763-6769.

91. Lazzari, R., IsGISAXS: a program for grazing-incidence small-angle X-ray scattering analysis of supported islands. *Journal of Applied Crystallography* **2002**, 35, 406-421.

92. Pedersen, J. S.; Vysckocil, P.; Schonfeld, B.; Kostorz, G., Small-angle neutron scattering of precipitates in Ni-rich Ni-Ti alloys. II. Methods for analyzing anisotropic scattering data. *Journal of Applied Crystallography* **1997**, 30, 975-984.

93. Dietrich, S.; Haase, A., Scattering of X-Rays and Neutrons at Interfaces. *Physics Reports-Review Section of Physics Letters* **1995**, 260, (1-2), 1-138.

94. Gutmann, J. S. Strukturbildung in dünnen Filmen aud Mischungen statistischer Copolymere. Johannes Gutenberg-Universität, Mainz, 2000.

95. Lazzari, R. IsGISAXS a program for analyzing Grazing Incidence Small Angle X-Ray Scattering from nanostructures. http://www.insp.upmc.fr/axe2/Oxydes/IsGISAXS/isgisaxs.htm

96. Auroy, P.; Auvray, L., Collapse-Stretching Transition for Polymer Brushes - Preferential Solvation. *Macromolecules* **1992**, 25, (16), 4134-4141.

97. Milner, S. T., Compressing Polymer Brushes - a Quantitative Comparison of Theory and Experiment. *Europhysics Letters* **1988**, 7, (8), 695-699.

98. Milner, S. T.; Witten, T. A.; Cates, M. E., Theory of the Grafted Polymer Brush *Macromolecules* **1988**, 21, (8), 2610-2619.

99. Degennes, P. G., Conformations of Polymers Attached to an Interface. *Macromolecules* **1980**, 13, (5), 1069-1075.

100. Luzinov, I.; Minko, S.; Tsukruk, V. V., Adaptive and responsive surfaces through controlled reorganization of interfacial polymer layers. *Progress in Polymer Science* **2004**, 29, (7), 635-698.

101. Schild, H. G., Poly (N-Isopropylacrylamide) - Experiment, Theory and Application. *Progress in Polymer Science* **1992**, 17, (2), 163-249.

102. Kato, K.; Uchida, E.; Kang, E. T.; Uyama, Y.; Ikada, Y., Polymer surface with graft chains. *Progress in Polymer Science* **2003**, 28, (2), 209-259.

References

103. Nath, N.; Chilkoti, A., Creating "Smart" surfaces using stimuli responsive polymers. *Advanced Materials* **2002,** 14, (17), 1243-+.

104. Schnell, R.; Stamm, M. In *The self-organization of diblock copolymers at polymer blend interfaces*, 1st European Conference on Neutron Scattering (ECNS 96), Interlaken, Switzerland, Oct 08-11, 1996; Elsevier Science Bv: Interlaken, Switzerland, 1996; pp 247-249.

105. Hsu, C. C.; Prausnit.Jm, Thermodynamics of Polymer Compatibility in Ternary-Systems. *Macromolecules* **1974,** 7, (3), 320-324.

106. Silberberg, A., Distribution of Conformations and Chain Ends near the Surface of a Melt of Linear Flexible Macromolecules. *Journal of Colloid and Interface Science* **1982,** 90, (1), 86-91.

107. Mukhopadhyay, M. K.; Jiao, X.; Lurio, L. B.; Jiang, Z.; Stark, J.; Sprung, M.; Narayanan, S.; Sandy, A. R.; Sinha, S. K., Thickness induced structural changes in polystyrene films. *Physical Review Letters* **2008,** 101, (11), 4.

108. Schweizer, K. S.; Curro, J. G., Integral equation theories of the structure, thermodynamics, and phase transitions of polymer fluids. In *Advances in Chemical Physics, Vol 98*, John Wiley & Sons Inc: New York, 1997; Vol. 98, pp 1-142.

109. Baschnagel, J.; Meyer, H.; Varnik, F.; Metzger, S.; Aichele, M.; Muller, M.; Binder, K., Computer simulations of polymers close to solid interfaces: Some selected topics. *Interface Science* **2003,** 11, (2), 159-173.

110. Keddie, J. L.; Jones, R. A. L.; Cory, R. A., Size-Dependent Depression of the Glass-Transition Temperature in Polymer-Films. *Europhysics Letters* **1994,** 27, (1), 59-64.

111. Forrest, J. A.; DalnokiVeress, K.; Dutcher, J. R., Interface and chain confinement effects on the glass transition temperature of thin polymer films. *Physical Review E* **1997,** 56, (5), 5705-5716.

112. Forrest, J. A.; DalnokiVeress, K.; Stevens, J. R.; Dutcher, J. R., Effect of free surfaces on the glass transition temperature of thin polymer films. *Physical Review Letters* **1996,** 77, (10), 2002-2005.

113. Ellison, C. J.; Torkelson, J. M., The distribution of glass-transition temperatures in nanoscopically confined glass formers. *Nature Materials* **2003,** 2, (10), 695-700.

114. Reich, S.; Cohen, Y., Phase-Separation of Polymer Blends in Thin-Films. *Journal of Polymer Science Part B-Polymer Physics* **1981,** 19, (8), 1255-1267.

References

115. Tanaka, K.; Yoon, J. S.; Takahara, A.; Kajiyama, T., Ultrathinning-Induced Surface Phase-Separation of Polystyrene Poly(Vinyl Methyl-Ether) Blend Film. *Macromolecules* **1995**, 28, (4), 934-938.

116. Ramakrishnan, A.; Dhamodharan, R.; Ruhe, J., Controlled growth of PMMA brushes on silicon surfaces at room temperature. *Macromolecular Rapid Communications* **2002**, 23, (10-11), 612-616.

117. Gianneli, M.; Roskamp, R. F.; Jonas, U.; Loppinet, B.; Fytas, G.; Knoll, W., Dynamics of swollen gel layers anchored to solid surfaces. *Soft Matter* **2008**, 4, (7), 1443-1447.

118. Guinier, A.; Fournet, G., *Small-Angle Scattering of X-rays*. Wiley: New York, 1955.

119. Porod, G., *Small-Angle X-ray Scattering*. Academic Press: London, 1982.

120. Memesa, M.; Weber, S.; Lenz, S.; Perlich, J.; Berger, R.; Muller-Buschbaum, P.; Gutmann, J. S., Integrated blocking layers for hybrid organic solar cells. *Energy Environ. Sci.* **2009**, 2, 783 - 790.

121. Frömsdorf, A.; Capek, R.; Roth, S. V., m-GISAXS experiment and simulation of a highly ordered model monolayer of PMMA-beads. *Journal of Physical Chemistry B* **2006**, 110, (31), 15166-15171.

122. Beaucage, G.; Rane, S.; Sukumaran, S.; Satkowski, M. M.; Schechtman, L. A.; Doi, Y., Persistence length of isotactic poly(hydroxy butyrate). *Macromolecules* **1997**, 30, (14), 4158-4162.

123. Chavan, S. V.; Sastry, P. U.; Tyagi, A. K., Deagglomeration and fractal behavior of Y2O3 nano-phase powders. *Scripta Materialia* **2006**, 55, (6), 569-572.

124. Tappan, B. C.; Huynh, M. H.; Hiskey, M. A.; Chavez, D. E.; Luther, E. P.; Mang, J. T.; Son, S. F., Ultralow-density nanostructured metal foams: Combustion synthesis, morphology, and composition. *Journal of the American Chemical Society* **2006**, 128, (20), 6589-6594.

125. di Stasio, S.; Mitchell, J. B. A.; LeGarrec, J. L.; Biennier, L.; Wulff, M., Synchrotron SAXS (in situ) identification of three different size modes for soot nanoparticles in a diffusion flame. *Carbon* **2006**, 44, (7), 1267-1279.

126. Norman, A. I.; Ho, D. L.; Karim, A.; Amis, E. J., Phase behavior of block co-poly (ethylene oxide-butylene oxide), E18B9 in water, by small angle neutron scattering. *Journal of Colloid and Interface Science* **2005**, 288, (1), 155-165.

References

127. Rathgeber, S.; Pakula, T.; Urban, V., Structure of star-burst dendrimers: A comparison between small angle x-ray scattering and computer simulation results. *Journal of Chemical Physics* **2004,** 121, (8), 3840-3853.

128. Higgins, J. S.; Benoit, H., *Polymers and Neutron Scattering*. Oxford, 1994.

129. Beaucage, G.; Kammler, H. K.; Pratsinis, S. E., Particle size distributions from small-angle scattering using global scattering functions. *Journal of Applied Crystallography* **2004,** 37, 523-535.

130. Ilavsky, J.; Jemian, P. R., Irena: tool suite for modeling and analysis of small-angle scattering. *Journal of Applied Crystallography* **2009,** 42, 347-353.

131. Fleer, G. J., *Polymers at interfaces*. Chapman & Hall: London [u.a.], 1998; p XIX, 495 S.

132. Zhao, B.; Brittain, W. J., Polymer brushes: surface-immobilized macromolecules. *Progress in Polymer Science* **2000,** 25, (5), 677-710.

133. Marciniec, B., *Comprehensive handbook on hydrosilation*. Pergamon Press: Oxford (u.a.), 1992; p XI,754 S.

134. Shuto, K.; Oishi, Y.; Kajiyama, T.; Han, C. C., Aggregation Structure of a 2-Dimensional Ultrathin Polystyrene Film Prepared by the Water Casting Method. *Macromolecules* **1993,** 26, (24), 6589-6594.

135. Parratt, L. G., Surface Studies of Solids by Total Reflection of X-Rays. *Physical Review* **1954,** 95, (2), 359-369.

136. Nevot, L.; Croce, P., Characterization of Surfaces by Grazing X-Ray Reflection - Application to Study of Polishing of Some Silicate-Glasses. *Revue De Physique Appliquee* **1980,** 15, (3), 761-779.

137. Dee, G. T.; Sauer, B. B., The Surface-Tension of Polymer Blends - Theory and Experiment. *Macromolecules* **1993,** 26, (11), 2771-2778.

138. Choi, K.; Jo, W. H.; Hsu, S. L., Determination of equation-of-state parameters by molecular simulations and calculation of the spinodal curve for polystyrene/poly(vinyl methyl ether) blends. *Macromolecules* **1998,** 31, (4), 1366-1372.

139. Brandrup, J.; Immergut, E. H.; Grulke, E. A., *Polymer handbook*. Wiley: New York, 1999.

140. Edmondson, S.; Osborne, V. L.; Huck, W. T. S., Polymer brushes via surface-initiated polymerizations. *Chemical Society Reviews* **2004,** 33, (1), 14-22.

141. Milner, S. T.; Witten, T. A.; Cates, M. E., Effects of Polydispersity in the End-Grafted Polymer Brush. *Macromolecules* **1989,** 22, (2), 853-861.

References

142. Hawker, C. J.; Bosman, A. W.; Harth, E., New polymer synthesis by nitroxide mediated living radical polymerizations. *Chemical Reviews* **2001,** 101, (12), 3661-3688.

143. Mayadunne, R. T. A.; Rizzardo, E.; Chiefari, J.; Krstina, J.; Moad, G.; Postma, A.; Thang, S. H., Living polymers by the use of trithiocarbonates as reversible addition-fragmentation chain transfer (RAFT) agents: ABA triblock copolymers by radical polymerization in two steps. *Macromolecules* **2000,** 33, (2), 243-245.

144. Moad, G.; Chiefari, J.; Chong, Y. K.; Krstina, J.; Mayadunne, R. T. A.; Postma, A.; Rizzardo, E.; Thang, S. H., Living free radical polymerization with reversible addition-fragmentation chain transfer (the life of RAFT). *Polymer International* **2000,** 49, (9), 993-1001.

145. Ejaz, M.; Yamamoto, S.; Ohno, K.; Tsujii, Y.; Fukuda, T., Controlled graft polymerization of methyl methacrylate on silicon substrate by the combined use of the Langmuir-Blodgett and atom transfer radical polymerization techniques. *Macromolecules* **1998,** 31, (17), 5934-5936.

146. von Werne, T.; Patten, T. E., Atom transfer radical polymerization from nanoparticles: A tool for the preparation of well-defined hybrid nanostructures and for understanding the chemistry of controlled/"living" radical polymerizations from surfaces. *Journal of the American Chemical Society* **2001,** 123, (31), 7497-7505.

147. Ramakrishnan, A.; Dhamodharan, R.; Jürgen, R., Controlled Growth of PMMA Brushes on Silicon Surfaces at Room Temperature. In 2002; Vol. 23, pp 612-616.

148. Urayama, K.; Yamamoto, S.; Tsujii, Y.; Fukuda, T.; Neher, D., Elastic properties of well-defined, high-density poly(methyl methacrylate) brushes studied by electromechanical interferometry. *Macromolecules* **2002,** 35, (25), 9459-9465.

149. Scattering Length Density Calculator. http://www.ncnr.nist.gov/resources/sldcalc.html

150. Descas, R.; Sommer, J. U.; Blumen, A., Grafted Polymer Chains Interacting with Substrates: Computer Simulations and scaling. *Macromolecular Theory and Simulations* **2008,** 17, (9), 429-453.

151. Ullman, R., Small-Angle Neutron-Scattering from Polymer Networks. *Journal of Chemical Physics* **1979,** 71, (1), 436-449.

152. Bueche, F., Viscosity of polymers in concentrated solution. *Journal of Chemical Physics* **1956,** 25, (2), 599-600.

153. Wohlfarth, C.; Wohlfarth, B.; Landolt, H.; Börnstein, R.; Martienssen, W.; Madelung, O., *Surface tension of pure liquids and binary liquid mixtures.* Springer: Berlin [u.a.], 1997.

References

154. Mullins, J., *Journal of Applied Polymer Science* **1959**, (2), 257.

155. Edwards, S. F.; Vilgis, T., The Effect of Entanglements in Rubber Elasticity. *Polymer* **1986,** 27, (4), 483-492.

156. Edwards, S. F.; Vilgis, T. A., The Tube Model-Theory of Rubber Elasticity. *Reports on Progress in Physics* **1988,** 51, (2), 243-297.

Publications

Peer reviewed articles

Memesa, M.; Weber, S.; Lenz, S.; Perlich, J.; Berger, R.; Muller-Buschbaum, P.; Gutmann, J. S., Integrated blocking layers for hybrid organic solar cells. *Energy Environ. Sci.* **2009**, 2, 783 - 790.

Lenz, S.; Bonini, M.; Nett, S.K.; Memesa, M.; Lechmann, M.C.; Emmerling, S.; Kappes, R.; Timmann, A.; Roth, S.V.; Gutmann, J.S.; Global scattering functions: A tool for Grazing Incidence Small Angle X-Ray Scattering (GISAXS) data analysis of low correlated lateral structures, *The European Physical Journal - Applied Physics* 2009, revised.

Lenz, S.; Nett, S.K.; Memesa, M.; Roskamp, R.F.; Timmann, A.; Roth, S.V.; Berger, R.; Gutmann, J.S.; Thermal response of surface grafted two-dimensional polystyrene (PS)/polyvinylmethylether (PVME) blend films, *Macromolecules* 2009, submitted.

Ochsmann, J. W.; Lenz, S.; Emmerling, S.G.J.; Kappes, R.S.; Nett, S.K.; Maria C. Lechmann, M.C.; Roth, S.V.; Gutmann J.S.; PS-*b*-PEO block copolymer thin films as structured reservoirs for nanoscale precipitation reactions, *Journal of Polymer Science, Part B:Polymer Physics* 2009, submitted.

Proceedings

Lenz, S., Nett, S., Memesa, M., Berger, R., Timmann, A., Roth, S.V., Gutmann, J.S.; Phasebehaviour of ultra-thin polymer films on arrays of Micromechanical Cantilevers (MCS) investigated with GISAXS; Annual HASYLAB Report 2007.

Lenz, S. and Gutmann, J.S.; Interfacial mobility and switching in polymeric brushes; Annual FRM II Report 2008.

Publications

Conference contributions

Sebastian Lenz and Jochen S. Gutmann; Phase behaviour and transition of polymer-brushes on Microcantilevers; Poster presentation at *DPG Frühjahrstagung* Regensburg 2007.

Sebastian Lenz, Sebastian K. Nett, Mine Memesa, Rüdiger Berger, Jochen S. Gutmann, Andreas Timmann, Stefan V. Roth; Phase behaviour and transition of polymer-brushes on Microcantilevers; Poster presentation at the 2nd *GISAXS Workshop May 2007*, Hamburg.

Sebastian Lenz, Sebastian K. Nett, Mine Memesa, Rüdiger Berger, Jochen S. Gutmann, Andreas Timmann, and Stephan V. Roth; Phase behaviour of ultra-thin polymer films grafted on Micro-Cantilever-Sensors (MCS); Talk at *DPG Frühjahrstagung* Berlin 2008.

Sebastian Lenz, Sebastian K. Nett, Mine Memesa, Ruediger Berger, Jochen S. Gutmann, Andreas Timmann, and Stephan V. Roth; Phase behaviour of ultra-thin polymer films grafted on Micro-Cantilever-Sensors (MCS); Poster presentation at the *International Workshop on Nanomechanical Cantilever Sensors* Mainz 2008.

S. Lenz, S.K. Nett, M. Memesa, A. Timmann, S.V. Roth, J.S. Gutmann; Global scattering functions: A tool for GISAXS analysis; Poster presentation at *HASYLAB User Meeting* Hamburg 2008.

Lenz S., Nett S., Lellig P., Berger R., Gutmann J.S.; Dynamic investigations of thin polymer brush films on micro cantilever sensor arrays; Poster presentation at *Frontiers in Polymer Science – International Symposium Celebrating the 50th Anniversary of the Journal Polymer* Mainz 2009.

Sebastian Lenz, Sebastian Nett, Sebastian Emmerling, Mine Memesa, Jochen S. Gutmann; Microfocus GISAXS Investigations of Nanomechanical Cantilever Arrays; Talk at *International Workshop on Nanomechanical Cantilever Sensors* Jeju, Korea 2009.

Publications

Sebastian Lenz, Adrian Ruehm, Rüdiger Berger, Jochen S. Gutmann; Stress Relaxation During Swelling of PMMA Brushes; Talk at *International Workshop on Nanomechanical Cantilever Sensors* Jeju, Korea 2009.

S. Lenz, M. Bonini, S.K. Nett, M. C. Lechmann, S.G.J. Emmerling, R.S. Kappes, M. Memesa, A. Timmann, S.V. Roth, J.S. Gutmann; Global scattering functions: A tool for Grazing Incidence Small Angle X-Ray Scattering (GISAXS) data analysis of low correlated lateral structures; Poster presentation at *GISAS 2009 – Satellite conference of SAS 2009*, Hamburg.

Sebastian Lenz, Adrian Rühm, Rüdiger Berger, Jochen S. Gutmann; Interfacial mobility and switching in polymeric brushes; Talk at the *FRM II User Meeting* München 2009.

I want morebooks!

Buy your books fast and straightforward online - at one of the world's fastest growing online book stores! Environmentally sound due to Print-on-Demand technologies.

Buy your books online at
www.get-morebooks.com

Kaufen Sie Ihre Bücher schnell und unkompliziert online – auf einer der am schnellsten wachsenden Buchhandelsplattformen weltweit! Dank Print-On-Demand umwelt- und ressourcenschonend produziert.

Bücher schneller online kaufen
www.morebooks.de

OmniScriptum Marketing DEU GmbH
Heinrich-Böcking-Str. 6-8
D - 66121 Saarbrücken
Telefax: +49 681 93 81 567-9

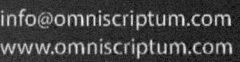

Printed by Books on Demand GmbH, Norderstedt / Germany